T0228036

Managing Food Security
in Unregulated Markets

Managing Food Security in Unregulated Markets

EDITED BY

Robert D. Reinsel

Routledge
Taylor & Francis Group

LONDON AND NEW YORK

First published 1993 by Westview Press

Published 2018 by Routledge
52 Vanderbilt Avenue, New York, NY 10017
2 Park Square, Milton Park, Abingdon, Oxon OX14 4RN

Routledge is an imprint of the Taylor & Francis Group, an informa business

Copyright © 1993 by Taylor & Francis

All rights reserved. No part of this book may be reprinted or reproduced or utilised in any form or by any electronic, mechanical, or other means, now known or hereafter invented, including photocopying and recording, or in any information storage or retrieval system, without permission in writing from the publishers.

Notice:
Product or corporate names may be trademarks or registered trademarks, and are used only for identification and explanation without intent to infringe.

Managing food security in unregulated markets / edited by Robert
    Reinsel.
        p.   cm.
    Includes bibliographical references.
    ISBN 0-8133-1704-5
    1. Food supply—Government policy.   2. Food industry and trade—
Government policy.   3. Grain trade—Government policy.
I. Reinsel, Robert D.
HD9000.6.M35   1993
338.1'9—dc20   92-38150
                                                                    CIP

ISBN 13: 978-0-367-00759-1 (hbk)

ISBN 13: 978-0-367-15746-3 (pbk)

Contents

Preface

The major grain producing nations are moving toward the reduction of domestic and export subsidies to agriculture. The grain importing nations are reducing import barriers. As world markets evolve, grain will tend to be produced in areas that have a comparative advantage in grain production. Over time, production will shift to least-cost areas.

Moving toward market orientation during the 1980's, the United States sharply modified its grain policy so that nonrecourse loans are no longer used as price enhancement devices. The loan rates are established at a percentage below the moving average price and now provide a safety net for prices when aggregate output is much larger than normal in relation to demand. This change tends to remove the United States from its long-term role as residual supplier to the world markets. U.S. grains are more likely to be priced competitively, and stocks are unlikely to accumulate in government storage.

The General Agreement on Trade and Tariffs (GATT), EC-92, the U.S. Farm Bill, and bilateral free trade agreements will likely further alter grain stocks policies. A reduction in U.S. grain stocks and a reduction in stock holding by other exporting nations implies that market supplies may more often be short and prices will more often rise sharply. Such shortages are likely to result in calls for an international buffer-stock system to alleviate the food security problems of developing countries. It has also been suggested that long-run prices will be higher with market oriented policies as total production declines relative to demand and export subsidies are eliminated or reduced.

Given these possible outcomes of policy change, the objective of this volume is to identify and explore a range of views on the possibilities for international stocks management as well as to determine what might be compatible with the free market. At the same time, consideration was given to protecting the interests of both importing and exporting nations.

This volume is thus concerned with: (1) the factors that give rise to differences in the commercial and government objective functions in stock holding, (2) the economic and social reasons for private and public stock holding, (3) alternative stock acquisition and dispersal programs (schemes), (4) the problems of increased supply and price variability, and (5) appropriate responses to the increased variability by developing and developed countries.

I begin the volume by setting the stage for the remainder of the book by showing how past experiences with free market prices in tight markets resulted in efforts to provide for more stable delivery of supplies and efforts to achieve price stability. Next, Alexander Sarris discusses the level of world cereal stocks and production variability in a more liberalized market environment. Peter Hazell follows with a chapter discussing the implications of grain trade liberalization for less developed country (LDC) food security. Don Gunasekera and Brian Fisher look at Australia's experience with the collapse of its wool buffer stock, including the critical factors leading to its collapse, the current wool stockpile, and its disposal. Don McClatchy and others summarize the Canadian experience with, and reliance on, income security; they also discuss income stability measures to deal with variability. Also, Takamas Akiyama suggests that there is a role for commodity futures in providing for LDC food security. In the final chapter, I discuss the causes and types of variability. I then turn to private and public sector stock and stability objectives and consider what adjustment methods might be compatible with the objective of maintaining market orientation and market pricing.

I also propose that the intra-year storage function is primarily the business of the commercial sector. Storage of planned production is a function of commercial markets. On the other hand, storage of unplanned production requires facilities that are not part of the commercial decision-making process. That is, it is not profitable to build facilities with the expectation that they will be needed in one or two years out of five. Storage above pipeline requirements appears to be a legitimate function of governments if the storage program is used for stabilizing market supply and prices and is not used for enhancing price.

Sarris suggests that agricultural trade liberalization has more complex implications for world price, production, and stock variabilities than has hitherto been thought. His first major conclusion was that stock adjustment is quite substantial in the current world cereal markets and would play an even more significant role in stabilizing world prices after trade liberalization. However, trend levels of world stocks are likely to fall after liberalization, as the public sector would lessen its interference in cereal markets, and the private sector would most likely not fully compensate for the reduction in publicly held stocks.

Sarris notes that developed market economies as a whole tend to insulate their domestic cereal markets more than the developing and centrally planned economies aggregated together, and that price transmission elasticities seem to be smaller for wheat and rice as compared with coarse grains.

Despite the substantial reduction in world price variability predicted under cereal trade liberalization, production is not greatly stabilized and in some products and regions it is even somewhat less stable. This is the result of short-run production response that now operates on relatively more variable domestic prices because of increases in price transmission. Changes in price and production variabilities have direct implications for world cereal stock levels. Average world cereal stock levels as a share of world production do not significantly change if only developed market economies (DMCs) liberalize and will fall somewhat if all countries liberalize. Under the parameter assumptions made, the world cereal stock levels, after complete trade liberalization by all countries might fall to levels below seventeen to eighteen percent of world production.

Hazell refers to a number of things that governments can do to reduce their country's exposure to catastrophic price risks. He suggests that these alternatives appear more promising than international solutions, and that we should not lose sight of them in focusing the discussion of this working group on international stocks. These options include diversifying production, emergency stock holding, reserve foreign exchange holding, and reducing production variability.

Gunasekera and Fisher suggest that Australia's recent experience with the collapse of its wool-price stabilization scheme provides several important lessons concerning stability programs. First, when the reserve price was conservatively set (particularly during the 1970's and the early part of the 1980's), it did little more than provide a safety net for growers under extreme circumstances. It is possible that some form of price stabilization might have been justified on economic efficiency grounds to compensate for failure in insurance and capital markets. However, available evidence indicates that in reality the operating costs of the scheme outweighed any insurance value generated. This conclusion is further reinforced once costs other than operating costs are taken into account, such as resource misallocation between competing farm industries. There is little evidence that the scheme was capable of stabilizing wool growers' incomes at the farm level. Further, there is no unequivocal evidence that the scheme stimulated world wool demand and may even have had adverse effects in that area. Therefore, there is little to suggest any gains from the scheme in terms of improved economic efficiency.

McClatchy and others point out that Canadian stabilization programs appear to be evolving from (a) full federal funding to (b) federal/producer cost sharing and finally to (c) federal/provincial/producer cost sharing. Also, many different provincial programs and national programs applied regionally are being consolidated to a few major cost-shared federal/provincial programs applied country wide. The evolution is from aggregate area program accounts to individual farm accounts; from a commodity focus to a whole farm income focus and from programs with potential deficits to programs with no deficits. Canada's experience with previous efforts to stabilize international grains prices through price band agreements and international stocks schemes as well as other methods have not been very positive. Past attempts to develop viable International Wheat Agreements in the post-World War II period and an International Grains Agreement under the GATT are seen in Canada as having been largely unsuccessful. International commodity agreements in general are viewed with skepticism. Furthermore, McClatchy et al. present the impression that buffer stocks programs tend to end up costing more than analysts forecast, possibly because a greater degree of omniscience on the part of the stabilizing authority is required than with underwriting or buffer fund schemes.

Akiyama and Trivedi evaluate a futures program for small developing countries. Limited research has been undertaken by using futures contracts for the purpose of stabilizing imported food prices. Akiyama and Trivedi concentrate on a variant of the competitive storage model. The analysis suggests that the futures program should fare better than the spot program, because the variance of prices is reduced when the futures prices are rationally expected prices, and hedging can be done before the supply disturbances are known to the market. While there are several unresolved research issues here, the available analyses suggest that a futures program should be useful in avoiding very high prices that occur occasionally in commodity markets.

In concluding the volume, I suggest that the objective of an independent stocks program would be to smooth the flow of unanticipated product to the market in response to unanticipated short supplies. If resources are committed with the expectation of normal yields and if prices and the output result in a significantly better or poorer crop, prices and incomes can be dramatically altered, even though the producers planned appropriately given their limited information. Neither the government nor the farmer can correctly anticipate or forecast the outcome of a specific crop at planting time, except by chance. Stocks programs, therefore, should react to crop output rather than anticipate crop output. Protecting farmers and consumers against random shocks to the system need not distort long-term market signals if the shocks are due to weather.

Robert D. Reinsel

About the Contributors

Takamas Akiyama is a Principal Economist in the International Trade Division, International Economics Department, The World Bank.

Brian S. Fisher is Director, Australian Bureau of Agricultural and Resource Economics.

J. Gellner is with the Policy Branch, Agriculture Canada.

B. Gilmour is with the Policy Branch, Agriculture Canada.

H. Don B. H. Gunasekera is Manager, International Economic Analysis Section, Australian Bureau of Agricultural and Resource Economics.

Peter Hazell is a Principal Economist with the Agricultural Policy Division, The World Bank.

Bruce Huff is with the Policy Branch, Agriculture Canada.

D. McClatchy is with the Policy Branch, Agriculture Canada.

Robert D. Reinsel is a Senior Economist with the Agriculture and Trade Analysis Division of the Economic Research Service, United States Department of Agriculture.

Alexander H. Sarris is professor of economics in the Department of Economics, University of Athens, Greece.

Pravin K. Trivedi, professor of economics at the Indiana University, Bloomington, Indiana, was a Visiting Research Fellow at The World Bank during the time his chapter was written.

1

International Stocks Management in Unregulated Markets

Robert D. Reinsel

The Issue

The major grain producing nations are moving, albeit slowly, toward the reduction of domestic and export subsidies to agriculture. The grain importing nations are reducing import barriers. World markets are evolving toward greater integration through free trade areas.

The General Agreement on Trade and Tariffs (GATT), EC-92, the 1990 U.S. Farm Bill, and multilateral free trade agreements can alter grain stocks policies sharply. A reduction in U.S. grain stocks and a reduction in stock holding by other exporting nations seems, potentially, to mean that markets will more often be short of grain and prices may more often rise sharply. Such shortages are likely to result in calls for an international buffer stock system to alleviate the food security problems of developing countries. Also, it has been suggested that long run prices will be higher in a market-oriented world as production, in the aggregate, is lessened relative to demand and export subsidies are eliminated or reduced.

The objective of this volume is to identify and explore a range of views on the possibilities for government intervention in a nondistorting manner in primarily unregulated markets.

Items appropriate for discussion include food security, production variability, demand growth, supply response, residual suppliers and buyers, buffering mechanisms, acquisition and dispersal rules for buffers, and cereal market strategies to deal with variability. The volume focuses on:

- the forces that create government objectives for providing income and price stability,
- appropriate responses to these forces given market orientation as an objective,
- the factors that give rise to the differences in the commercial and government objective functions in stock holding,
- the economic and social reasons for private and public stock holding,
- alternative stock acquisition and dispersal programs (schemes), and
- the problems of increased supply and price variability, and appropriate responses to the increased variability by developing and developed countries.

I set the stage for the following chapters showing how past experiences with free market prices in tight markets result in efforts to provide for more stable delivery of supplies and efforts to achieve price stability.

In Chapter two Alexander Sarris discusses the level of world cereal stocks and production variability in a more liberalized market environment.

In Chapter three, Peter Hazell deals with the implications of grain trade liberalization for less developed countries (LDC) food security.

Next, Don Gunasekera and Brian Fisher review Australia's experience with the collapse of its wool buffer stock. Issues covered will include the factors that led to the collapse of the wool buffer stock, the size of the current wool stockpile, and the disposal of the stockpile.

In chapter five, Don McClatchy discusses the Canadian experience with, and reliance, on income security and income stability measures.

Also, Takamas Akiyama illustrates the use of commodity futures and options for food security. Emphasis is given to developing economies as importers.

I then close by discussing causes of variability and private and public sector stock and stability objectives and consider what methods of adjustment are compatible with the objective of maintaining market pricing.

Agricultural Policy and Grain Market Stability

In the 1930's, 1940's and 1950's major producers attempted to develop international agreements to share the wheat market and also attempted to prevent sharp price declines in periods of excess supplies.

Twice since 1970 world-wide concern has been expressed about food security in the developing countries. In both periods, world grain stocks had declined to near pipeline levels and grain prices were rising, leading to concerns over both short-run and long-run availability of food. These concerns resulted in calls for buffer stocks to prevent sharp price increases.

Neither the market agreement nor buffer stock schemes were successful. National interests broke the wheat agreement and prevented the international buffer stock from being developed.

The price support policies of the United States, in the 1950's and 1960's, cause this country to provide the world market supply buffer for grains. With support prices above the world market price in most years, grain tended to accumulate in government supported storage. Grain moved onto the market only when prices were above the loan rate or through subsidized exports.

From 1972 to 1979, markets tended to clear above the U.S. support level. However, from 1979 to 1986, grain tended to accumulate in the U.S. Farmer Owned Reserve (FOR), which provided long-term nonrecourse loans and storage subsidies. Because release prices were high, the program tended to make price fluctuations in the world market very large.

Since 1986, the U.S. agricultural policy has changed so that it no longer acquires and holds grain stocks to keep grain prices above longer run free market levels. The United States no longer provides the supply buffer for the international market. U.S. grain is priced competitively and grain stocks are less likely to accumulate under nonrecourse loans or in government storage.

On an international scale, governments are working to remove policy distortions from agricultural markets. The GATT, EC-92, the North American Free Trade Agreement (NAFTA), and other efforts at international cooperation hold the promise for less protectionism and a more efficient flow of commodities in trade. Yet, free markets will not necessarily have more stable prices.

Instability in grain markets is primarily the result of supply shocks. Demand by countries is nearly a constant, with only slight responsiveness to income and prices. Aggregate demand grows primarily in response to growth in population. Free trade will tend to overcome the problem of instability only to the extent that quotas, tariffs, and subsidies create barriers to trade and destabilize markets. However, supply instability caused by yield shocks from pests or weather will continue to be a major force driving prices, and a free competitive markets will not, of itself, remove the instability. Weather shocks are not caused by economic forces. Nothing was invested to achieve such a yield increase or decrease. Such changes are noneconomic; that is, they had no basis in economic decisions and do not respond to economic signals.

The question arises, then, whether the commercial sector can accommodate inter-year transfer of grain supplies to remove the impact of the supply shock or whether government intervention will be required?

Commercial Storage

Commercial firms are assumed to have profit maximization as their objective function. They tend, like their owners, to be risk averse. They plan according to the most probable outcome. And their most likely or most probable outcome is normal yield. There are no profits to be made in continuously expecting low or high crop yields.

Commercial firms will tend to maximize profits by planning to carry only pipeline needs to the next season. They have very little incentive for carrying large inter-year stockpiles of grain, particularly when they can pass the higher costs along to consumers.

Previous Analysis

The Food and Agriculture Organization of the United Nations (FAO) published a report, *Approaches to World Food Security (1)*, in 1983 in which it expressed concern over the longer term availability of grain stocks. The report suggests possible actions to assist the developing nations in dealing with the effects of grain shortages.

In the report, FAO defines three types of food shortages: (1) A "demand gap" that is equal to demand (consumption) minus production, (2) A "physiological gap" that is equal to nutritional requirements minus production and imports, and (3) An "emergency gap" that is equal to planned production and imports minus actual production and imports.

The three gaps provide a useful basis for exploring the causes of grain shortages and surpluses; the need for storage; and the methods of acquiring, holding, and disposing of stocks.

The Demand Gap

A demand gap, as defined by FAO, means that a country is not self-sufficient in the production of a commodity. Its population wants to consume more of the commodity than it can produce under normal conditions.

Demand gaps are more or less permanent conditions. Given the price that prevails in the market, the country finds it less expensive to import the commodity than to produce it. The country may not have the appropriate resources to produce the commodity. Or, it may be that its costs of production are higher than transportation plus production costs in other countries.

A demand gap may also occur because of government policy. Taxes on production or subsidies to consumption can create supply or demand shifts that produce a demand gap, or, alternatively, a surplus.

Demand gaps are filled by allowing imports of the commodity, by allowing prices to rise, or by removing policy barriers to production. If the country has sufficient foreign exchange or commodities for barter, it can close the demand gap with imports.

In a stable world, demand gaps create specialization and trade. World production increases because each nation uses its resources to produce those things at which it is most efficient.

The Physiological Gap

A physiological gap occurs because some members of the society are unable to acquire sufficient nutrients through their own production or through the market to meet their physiological needs. They are not part of the effective demand. Because they lack the resources to acquire the food needed to sustain life, they slowly become more and more debilitated and eventually starve.

Physiological gaps are longrun problems that occur because of a mismatch between resources and population. The members of society do not possess the resources to feed themselves on a subsistence basis or to provide goods to the market in exchange for food and shelter. Although a market may exist, the ability of individuals to enter into the market is nonexistent or limited.

A short-run solution to a physiological gap is to deal with it through direct consumer subsidies with free or lower cost food provided by donor countries.

A longer term solution requires that the productivity of the individuals and the country increase so that individuals acquire the resources to either increase output of the basic food commodities or to produce goods for trade. Alternatively, immigration will allow the population to be reduced so that a better distribution of resources is achieved.

The Emergency Gap

As defined by FAO, an emergency gap is produced by an unplanned reduction in output through the effects of weather (drought, frost, rain), insects, or disease, or a combination of the three on yields. The yield shock reduces production below the planned or normal level. Planned consumption cannot be met from current production and planned imports at the planning price. Shortages occur in the market. Unless stocks are available, prices rise sharply, assuming that they are not regulated, and people and countries are crowded out of the market. Consumption declines and, if alternatives are not available, famine occurs. Alternatively, yield shocks can increase output dramatically, flooding the market with product. Prices can fall sharply and farmers' incomes decline so that losses can result in failure of the business.

Agricultural Policy and Food Security

The FAO report identifies five issues or questions to be answered about food security and market stability:

- Are emergency food reserves necessary at the world level?
- Are stabilization stocks required over and above normal stocks?
- Should stabilization stocks be under international control?
- How would the cost of stockholding policies compare with their expected benefits?
- How should reserves be acquired and released?

The remainder of our discussion during these four sessions will address these questions. Before turning the program over to Alex Sarris for his discussion, I will present the results of an analysis of the effects of weather and other stochastic shocks on grain yields and grain stocks.

Estimating the Effects of Yield Variability

To understand the effects of weather shocks and pest shocks on yield, these aberrations must be separated from the effects of technologies that induce long-term upward or downward trends in production per acre. For this analysis, a 15-year linear trend was assumed to represent an estimate of the effect of changes in the level of inputs and changes in the genetics of the crop. The 15-year trend was used to develop an estimate of the expected yield for the sixteenth year. Differences from the expected yield are assumed to represent the effects of weather and pests.

Fifteen-year trends were estimated for data from 1960 to 1990 for the major producing countries for wheat, corn, and barley. The per-acre differences between actual yields and expected yields were multiplied times the number of acres harvested to determine the aggregate yield shock for the country. These aggregates were summed to determine the global effect of yield on production. Changes in production not represented by the yield shock are assumed to be planned changes in acreage, technology, or policy and, therefore, due to economic forces that should be reflected in market prices. The weather-induced shock is not the result of economic decision-making. Removing such a shock from the market would improve market efficiency without distorting prices.

Wheat

The 10 major producers of wheat in this study are: the United States, the Soviet Union, Australia, Canada, EC-12, Argentina, China, India, Pakistan, and Turkey. They produce 85 to 90 percent of all wheat (Fig. 1.1).

Using the projected trend estimation procedure for wheat for the period from 1975 through 1990 the global yield shock ranged from 40 million metric tons below normal to 38 million metric tons above normal, or from a 11.5 percent loss to an 8 percent increase (Fig. 1.2). In the absence of buffer stocks or

Figure 1.1 Share of wheat stocks held by major producers
Percent

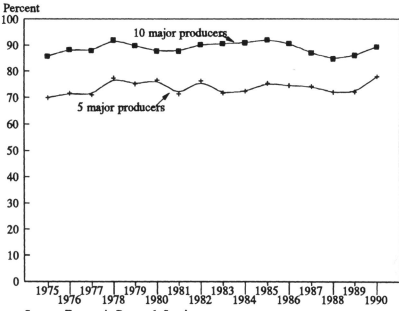

Source: Economic Research Service

Figure 1.2 Wheat stocks and yield shocks
Million metric tons

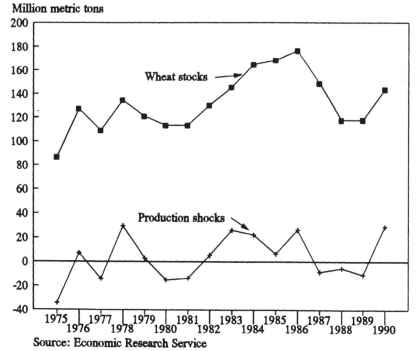

Source: Economic Research Service

Figure 1.3 Wheat stocks and production shocks

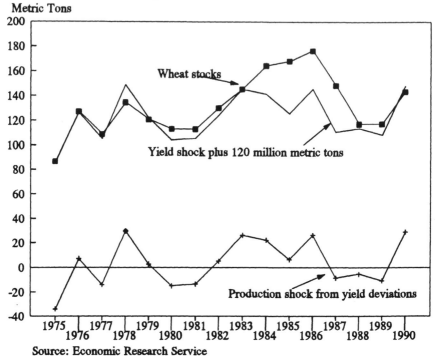

Source: Economic Research Service

border measures such quantity changes could result in sharp price increases (25 to 30 percent) above the expected price and price declines (16 to 20 percent). The percentage depends on the global elasticity of demand for wheat and the cross-price elasticity of substitutes.

A comparison of the global yield induced production changes with the stocks held by the major producers shows a very close correlation of actual ending stocks with yield deviation as estimated above. In fact, by adding a constant 120 million metric tons to the yield shock line in Figure 1.3 it can be superimpose over the stocks line producing an almost exact correspondence of the two lines (Fig. 1.3). A difference occurs during the 1983 to 1987 period when the 1981 U.S. Agricultural Act dramatically increased U. S. stocks.

The share of stocks held by each of the major producers varied sharply during the 1975-1990 period but the 10 producers consistently held a buffer of about 120 million metric tons above the yield shock.

A comparison of the yield shocks with the stocks shows that, if the stocks were readily available to the market, global carryover in the neighborhood of 60 million metric tons would have been sufficient to provide a stable market supply. However, stocks in some countries are held for domestic food security reasons only and are not really a part of the supply available to world trade.

It appears that the major producers on a global basis store and dispose of gain stocks in response to changes in yield deviations. But this may be more accidental than planned.

Corn

Seven producers including Argentina, Brazil, China, the EC-12, the Soviet Union, South Africa, and the United States raise 300 to 350 million metric tons, or about 75 to 80 percent of the world's corn (maize) (Fig. 1.4). And, they hold the majority of corn stocks, with the United States dominating the stock holding.

Stock holding by these countries tends to increase and decrease in line with yield shocks as estimated by the trend procedure, but not as closely as in the case of wheat. The divergence is primarily a function of U.S. and Chinese stock policies (Fig. 1.5 and Fig. 1.6).

Barley

Eight producers raise 75 to 80 percent of the world's barley and hold a proportional amount of the barley carryover stocks (Fig. 1.7). These countries are Canada, China, the United States, the EC-12, the Soviet Union, Poland, Iran, and Turkey. The largest producer's the EC-12 and the Soviet Union contribute between 55 and 60 percent of the total production.

Barley production changes based on yield shocks are dramatically affected by the production shocks in the Soviet Union the major producer of barley (Fig. 1.8). The production shock for barley is not as closely related to the change in stocks as with wheat or corn (Fig. 1.9). In part this reflects the change in the amount of grain fed to livestock as well as changes in the stock level.

Comparing wheat, corn, and barley stocks relative to the degree of production variation, one finds that wheat stocks are much higher relative to the production variation. This may result from concerns over food security and an unwillingness to adjust the level of human consumption.

Feed grains, such as corn or barley, represent much less of a concern for the human diet than wheat. Also, changes in the level of grain fed to livestock are much less of a threat to the food supply than changes in the availability of wheat.

The shocks imposed on markets by yield changes can have rather dramatic effects on world prices because demand of grains is inelastic with respect to prices. Such price shocks could have destabilizing effects on governments if consumers find the price of staple diet items rising sharply.

For commodities affected by weather and seasonal production patterns, the intra-year storage function is primarily the business of the commercial sector.

10

Figure 1.4 World corn production
Million metric tons

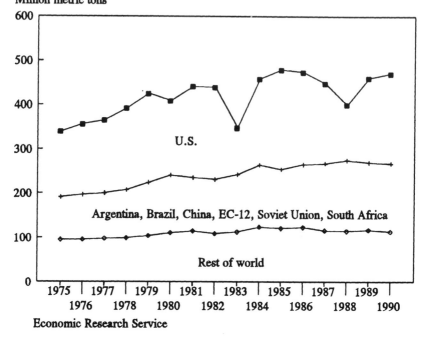

Economic Research Service

Figure 1.5 World corn stocks, and production shocks from yield deviations
Million metric tons

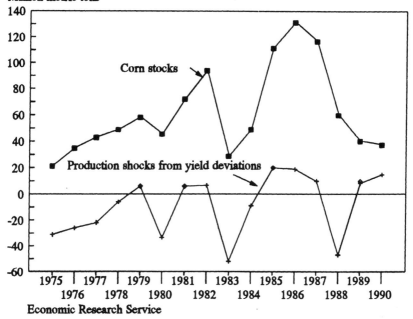

Economic Research Service

Figure 1.6 World corn stocks and production shocks from yield deviations
Million metric tons

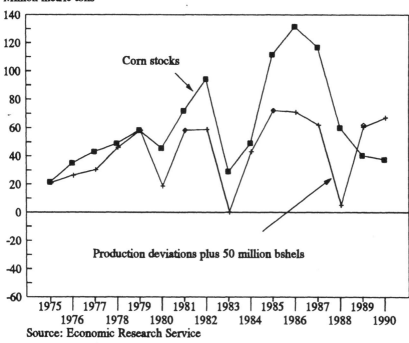

Source: Economic Research Service

Figure 1.7 World barley production
Million metric tons

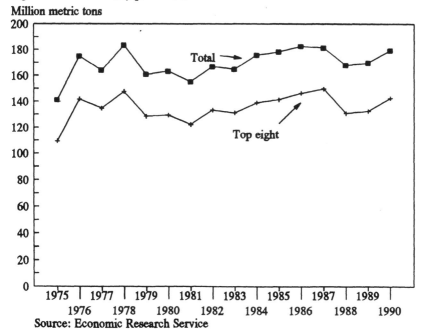

Source: Economic Research Service

Figure 1.8 World barley production shocks from yield deviations
Million metric tons

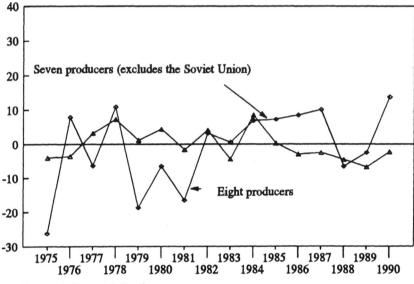

Economic Research Service

Figure 1.9 World barley stocks and production shocks from yield deviations
Million metric tons

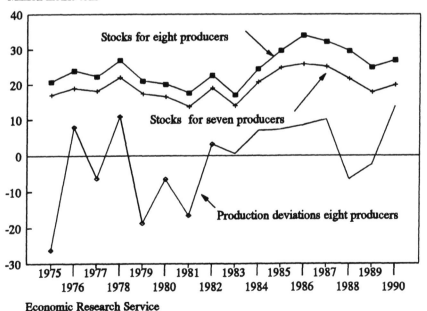

Economic Research Service

sector. Storage of planned production is a function of commercial markets. On the other hand, storage of unplanned production require facilities that are not part of the commercial decision-making process. That is it is not profitable to build facilities with the expectation that they will be needed 1 or 2 years out of 5.

Providing inter-year storage above pipeline needs to carry over the unplanned yield shock appears to be a legitimate function for governments to provide or subsidize if the storage program is for stabilizing market supply and prices and not for price enhancement.

References

FAO, *Approaches to World Food Security*, FAO Economic and Social Development No. 32, Commodities and Trade Division, Rome, Italy, 1983.

FAO, *Impact of National Grain Policies on World Grain Supplies and Prices*, FAO Economic and Social Development paper No. 56, Commodities and Trade Division, Rome, Italy, 1985.

FAO, *The Effect of Trade Liberalization on Cereal Stocks*, Committee on Commodity Problems, Intergovernmental Group on Grains, CCP:GR 90/3, August 1990.

Sarris, A., *The Impact of Agricultural Trade Liberalization on World Cereal Instability and Stocks*, Report prepared for FAO Commodities Division. Document ESC/M/90/1, July 1990.

2

Cereal Stocks and Production Variability in a Liberalized World Trade Environment

Alexander H. Sarris

Introduction

Despite the failure of the Uruguay Round of Multilateral Trade Negotiations (MTN's) to produce an accord on a framework for liberalizing agricultural trade, the consensus seems to be that the world slowly but steadily will move toward a more liberal agricultural trading environment. A more liberal world trade environment in agricultural products, and especially cereals (which form the bulk of world agricultural trade), implies that public interventions in stabilization operations and, hence, publicly held stocks will diminish. This raises the possibility, feared by many developing food importing countries, that international prices will become more unstable, but, more importantly, that the probability of large positive price spikes will become larger. This latter phenomenon, in turn, is related to the average level of stocks held around the world. It is reasonable to assume that lower average world stock levels will imply a much larger price increase in the event of a world production shortfall. While all empirical investigations of world agricultural trade liberalization imply a lower degree of world price instability, this does note rule out a more infrequent but more severe price spike.

While much discussion on world cereal instability has centered on the behavior of reserve stocks, normal working stocks form the bulk of total private stocks. It is the thesis of this chapter that the behavior of these so called "normal working" stocks under a more liberal environment will largely determine the overall level of world cereal stocks. The level of these stocks, in turn, is most likely influenced by variables such as production variability that are likely to shift under a more liberalized environment.

First, a general theoretical model of national excess supply is outlined and then the model is used to discuss trend levels of private stocks. Next, an equation for trend prices and private cereal stocks is derived and a model explaining the deviations of prices, production and stocks is outlined. The model is empirically specified and the empirical results of simulations of trade liberalization in wheat, rice, and coarse grains are outlined.

A General Theoretical Model of National Cereal Excess Supply

In this section, a general one-commodity model of excess supply for a cereal commodity is illustrated.

Denote by Q_t the production of the commodity in year t in some country i (the i will be suppressed for the time being), by D_t the domestic consumption of the commodity in the same year, by NX_t the net exports (namely exports minus imports) of the commodity, by SP_t the amount of stocks held by private agents at the end of period t, and by SG_t the end of period publicly held stocks. Then the supply demand balance for the commodity in that country and for year t is as follows:

$$Q_t + SP_{t-1} + SG_{t-1} = D_t + NX_t + SP_t + SG_t \tag{2.1}$$

By rearranging terms in (2.1) the following obtains:

$$NX_t + \Delta SG_t = Q_t - D_t - \Delta SP_t \tag{2.2}$$

where Δ is the difference operator ($\Delta x_t = x_t - x_{t-1}$).

The specification of behavioral rules for Q_t, D_t, and ΔSP_t clearly gives a definite expression for the right-hand side of (2.2), which specifies, in turn, the sum of the excess supply plus the variation in publicly held stocks. The level variables in the right-hand side of (2.2) will be specified as equal to trend levels plus deviations from trends as follows:

$$D_t = D_t^*(P_t^*, Z_t^*) - a(P_t - P_t^*) + u_t \tag{2.3}$$

$$Q_t = Q_t^*(P_t^*, Z_t^*) + b(P_{t/t-1}^e - P_t^*) + c(P_t - P_{t/t-1}^e) + v_t \tag{2.4}$$

$$SP_t = SP_t^*(Q_t^*, Z_t^*) + \alpha(P_{t+1/t}^e - P_t) - \beta SG_t. \tag{2.5}$$

In the above equations, variables with an asterisk as a superscript denote trend values and the other variables are defined as follows:

P_t =   level of real price of the commodity (namely, nominal price deflated by an index of prices of other goods)

Z^*_t =   a vector of trend variables driving supply and demand that are exogenous to the particular commodity market (technology, income, population, interest rates, etc.)

$P^e_{t/t-1}$ =   expectation of the value P_t given information, namely, realizations of random variables, up to and including period t-1

u_t, v_t = random terms affecting domestic demand and production, respectively.

Notice that there are already several salient features of this specification. First, trends are defined as longrun expected values; namely, expectations over all random variables in the system past and present. By contrast, the conditional expectations are based on information and, hence, on specific realizations of the past random variables. This implies that the trend of $P^e_{t/t-1}$ is equal to the longrun trend P^*_t.

Another salient implication of the above specification is that the random terms u_t and v_t include the impacts of deviation from trends of all exogenous variables Z_t. In Z_t, cross-commodity and general equilibrium effects are subsumed. This, for instance, explains why it is necessary to include a random term for domestic demand. It also implies that the random terms u_t and v_t might not be independent, as fluctuations in macroeconomic variables such as exchange rates, interest rates, etc. might affect both supply and demand of the commodity. This is almost assured in poor developing countries where production in key cereals significantly affects income and, hence, demand.

In equation (2.4), notice that both short-term (namely, within the production period) and lagged response to price are modeled. The coefficient b denotes the response of planned production to expectations held in the previous year. This is appropriate in a model of a commodity with an annual cycle. The coefficient c, in turn, denotes shortrun supply response. In practice, both area and yield can have lagged as well as short run responses. Since modelling both responses simultaneously would complicate the model, it is chosen to model total production instead. The above reasons, as well as the previous discussion about what is included in the random term v_t, also imply that there is no a-priori reason for multiplicative instead of additive uncertainty. I chose the latter specification, as it is analytically simpler.

Notice that in the stock specification, apart from the current response to expected price changes, a response to public stocks is included. If private stocks fully counteract public stock building, then the value of β should be equal to one. In other words, public stocks would fully crowd out private ones.

Otherwise, and as is most likely, β should be smaller than one. This is, in fact, what is suggested by theory (e.g., Newberry and Stiglitz, 1981).

Notice that I have not talked about "trend" public stocks. If, in fact, public stock building is an ongoing activity and the private sector has formed longrun expectations about the average level of public stocks, then equation (2.5) should be modified as follows:

$$SP_t = SP_t^* \ (Q_t^*, Z_t^*) - \beta SG_t^* + \alpha(P_{t+1/t}^e - P_t) - \gamma(SG_t - SG_t^*). \tag{2.6}$$

In (2.6), the trend level of private stocks (what will be termed later as the normal private carryover stocks), is influenced (negatively) by the trend level of public stocks via the parameter β. In the shortrun, however, the deviation of private stocks from trend is expected to be influenced by the short run behavior of public stocks. The point is that the response of trend levels of private stocks might be different than the short-term response and this point is illustrated by the inclusion of different response parameters β and γ.

The events preceding the 1973-74 food crisis suggest that a specification like (2.6) is more appropriate. It is quite likely that before 1973 the private sector in the United States and in other countries believed that the U.S. would always hold cereal stocks. In terms of the model above, this implies a nonzero value of SG_t^*, and a lowering of privately held "normal" stocks. In fact, if SG_t^* was perceived as very large, then this probably implies that the value of γ was quite small; namely, short run variations in public stocks did not matter much. The panic of 1973-74 could then be attributed to a sudden change in longrun expectation about the size of SG^* and a scramble to rebuild normal precautionary inventory levels. This would explain why, for instance, agents kept stockpiling in spite of the fact that prices kept rising.

The above discussion suggests that the notion of normal private carryovers should be modified to include longrun expectations about public stocks. In the sequel, the specification (2.6) will be kept.

Taking the longrun expected value of (2.3), (2.4,) and (2.6) an equation for the trends obtains:

$$NX_t^* + \Delta SG_t^* = Q_t^*(P_t^*, Z_t^*) - D_t^*(P_t^*, Z_t^*) - \Delta SP_t^*(Q_t^*, Z_t^*) + \beta \Delta SG_t^* \tag{2.7}$$

or

$$NX_t^* = Q_t^*(P_t^*, Z_t^*) - D_t^*(P_t^*, Z_t^*) - \Delta SP_t^*(Q_t^*, Z_t^*) - (1-\beta)\Delta SG_t^*. \tag{2.8}$$

Notice that the above specification is not at all trivial, because it includes trends in normal private carryovers (as well as trends in public stocks) in addition to normal production and consumption trends. This trend specification involving stock trends has been largely ignored in almost all previous theoretical and empirical investigations of agricultural trade liberalization (e.g., Roningen and Dixit, 1989; Burniaux et al., 1989; Parikh et al., 1988; and Tyers and Anderson, 1986).

If (2.8) is subtracted from (2.2) and equations (2.3), (2.4), and (2.6) are applied then the following shortrun equation is obtained:

$$nx_t = (\alpha + a + c)p_t - \alpha p_{t-1} - \alpha p^e_{t+1/t} + (\alpha + b - c)p^e_{t/t-1} - (1-\gamma)\Delta sg_t + v_t - u_t. \qquad (2.9)$$

Implicit in (2.9) is the assumption that the market price p_t (p_t is the deviation of market price from trend, but I use the term "market price" to save words) is the same for both producers and consumers. While the effective prices to producers and consumers are indeed different in most countries, the interest here lies mostly in the departures of domestic prices from international prices. In empirical applications, the distinction, of course, can easily be made. In equation (2.9), lower case letters denote deviations of upper case variables from trends. Notice that the deviations of current production and consumption from their trends include not only purely random terms (at least for this market) such as v_t and u_t, but also shortrun responses to price. Versions of models similar to (2.9) have been used in almost all prior theoretical analyses of price and production fluctuations.

Trend Levels of Private Stocks

While substantial analysis in the past has been done concerning trend levels of cereal prices, very little analysis has been done concerning the determination of trend stock levels. Most of the literature on commodity stocks has concerned the behavior of short-term speculative stocks. Implicit in these works is the notion that the level of normal inventories required to smooth out variations in demand of products processed from the primary commodity is a relatively constant fraction of production or sales, while deviations of actual stocks from this underlying "desired" level are due either to speculative motives or to unanticipated shortfalls or excesses in production and sales from forecasted levels.

It is a long time since it has been recognized that private commodity stocks that are willingly held are composed of three components: one for transactions purposes, another for precaution against stock-outs, and a third for speculative reasons (for an early survey of the relevant theory which goes back to Keynes and Working, see Weymar, 1968; for a more recent survey, see Ghosh et al., 1987). A fourth reason for holding stocks, unwillingly this time, is because forecasts turned out to be wrong and, hence, actual stocks are different than what was planned. The stocks held for transactions and precautionary motives are the same in concept as those the Food and Agriculture Organization (FAO, 1983) classified, in the context of food security discussions, as normal working plus reserve stocks.

The first two components of desired private stocks are held for what is known as convenience yield, a term coined by Kaldor (1939). The total convenience yield, denoted by $Y(SP)$, can be thought of as the total benefit to a holder of the commodity from holding a given amount, SP. The "marginal convenience yield" (denoted by y(SP)) obtained from holding an extra unit of the commodity as stock (given that the agent already holds a total amount SP) is simply the derivative of the total convenience yield with respect to SP. The well known properties of the marginal convenience yield (Weymar, 1968), are that $y(0) = \infty, y' < 0$ and $y'' > 0$ (where $(/)$ denotes the derivative), and, that as SP tends to become large, y(SP) tends to zero. Furthermore, the shape of y(SP) depends on the underlying probability distribution function of supply (or demand). The speculative demand for stocks, in turn, depends on expected price appreciation.

The upshot of the theory is that the desired level of end-of-period total private stocks will be such as to equate the expected monetary return from holding an extra unit of the commodity to the marginal cost of holding an extra unit of the commodity. The marginal cost of storage is equal to the physical storage cost minus the marginal convenience yield. In other words, the following implicit equation in SP_t will give the total desired level of private stocks:

$$P_{t+1/t}^e - (1+r)P_t = \zeta(SP_t) - y(SP_t) \tag{3.1}$$

where r is a rate of discount and $\zeta(SP_t)$ is a term denoting physical storage cost, with the property that $\zeta' > 0$ and $\lim \zeta(SP) = \infty$ as SP tends to infinity.

The solution to the problem is illustrated in Figure 2.1. The curve labeled AA is simply a graphic representation of (3.1). "Trend" or "normal" private carryovers can now be illustrated. According to the logic that led to the derivation of (2.7), if all random variables are set to their unconditional expectations in equation (3.1), then an equation implicit in SP_t will be obtained. The solution is illustrated in Figure 2.1 by SP_t^*. It is this level that will be referred to as the trend or normal carryover private stock. Note that this stock includes only the transactions and precautionary stock components (or normal working stocks plus reserve stocks according to FAO terminology), since the average level of speculative stocks is zero.

A point of the theory not usually discussed by economists, but well known to operations researchers dealing with inventory control, is that the shape of the marginal convenience yield is affected by the underlying probability distributions. This point can be easily visualized by referring to Figure 2.2. Figure 2.2 shows two different probability distributions of demand for the product of a given agent, who transforms the commodity in some way. Distributions A and B both have the same means, but A has smaller variance than B. The agent, if faced with distribution A, will opt to hold some level of stocks S_1 to cover unanticipated variations in sales. If, however, the sales distribution is B, then he will opt for more stocks S_2 as the uncertainty and, hence, his stock-out risk have increased. Going back to Figure 2.1, an increase in the variability of supply or demand should lead to a rightward shift of the

Figure 2.1 Determination of end of period private stocks

$$P_{t+1_t}^o - (1+r)P_t$$

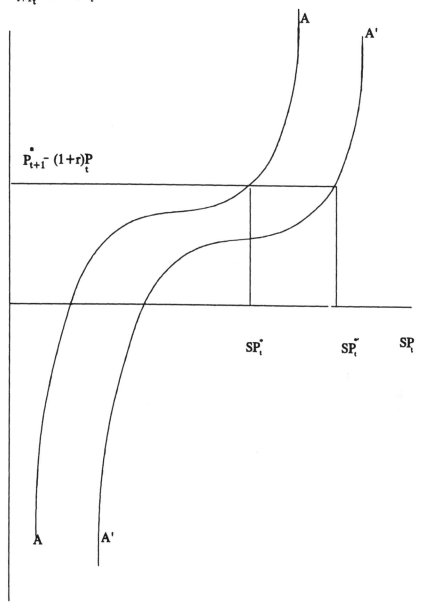

22

Figure 2.2 Private precautionary stocks

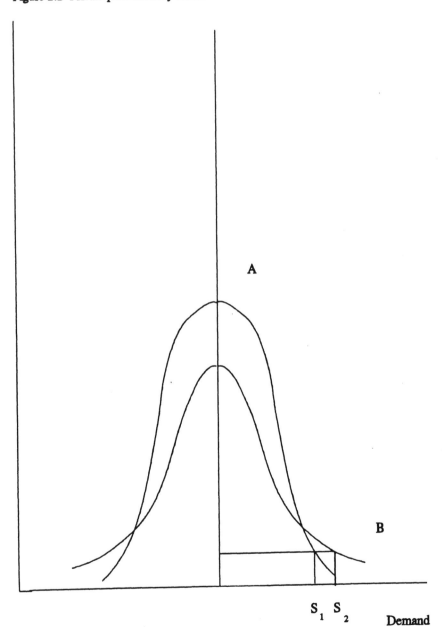

A

B

S$_1$ S$_2$

Demand

whole curve AA to $A'A'$, and, hence, will tend to increase *ceteris paribus* the level of "normal" end-of-season stocks.

The points of the previous discussion are, first, that SP^*_t should be a function of the underlying uncertainty. Second, trade liberalization in a cereal commodity market might change the underlying uncertainty faced by agents, thus leading to changes in the trend or "normal" component of total private stocks.

A Model of Trend Prices and Private Cereal Stocks

Given the discussion of the previous section, an analysis of trend levels of prices and private stocks can be obtained. The starting point will be equation (2.8), which indicates the trend levels of the market variables. The following specifications of trend production, consumption, and desired private stocks will now be made:

$$Q^*_t(P^*_t,Z^*_t)=\overline{Q}_t+dP^*_t \qquad (4.1)$$

$$D^*_t(P^*_t,Z^*_t)=\overline{D}_t-eP^*_t \qquad (4.2)$$

$$SP^*_t(Q^*_t,Z^*_t)=(\delta+\varepsilon CV)Q^*_t=(\delta+\varepsilon CV)\ (\overline{Q}_t+dP^*_t). \qquad (4.3)$$

In (4.1) and (4.2), $\overline{Q}_t$ and $\overline{D}_t$ denote the production consumption trends that are associated with variables other than trend price (e.g., macro variables and substitution effects among commodities). In (4.3), CV denotes the coefficient of variation of production which will be assumed constant under a given trade regime. Given standard inventory theory, it would appear that the coefficient of variation of demand (as a proxy for sales) would be the variable most suitable for inclusion in (4.3). However, in annual agricultural commodity markets such as cereals, the major source of uncertainty and the major reason private agents hold precautionary stocks is variability of supply. Nevertheless, one could include the coefficient of variation of demand as an additional term in the coefficient multiplying trend production in (4.3). This is not done here for simplicity. The underlying theory behind (4.3) is a simple variable accelerator model of desired inventory behavior, the accelerator being the coefficient multiplying trend production in (4.3). This is naturally assumed to be positive and smaller than one. Note that the trend or the "normal" level of private stocks, depends on two variables which will change under trade liberalization. These are the coefficient of variation of production and the trend price.

Given the discussion in the previous section, one might want to think of the parameter δ in (4.3) as the share of total trend private stocks held for transaction purposes. One might also call this, following FAO terminology, the "normal working" stock. The other term in (4.3), ϵCV, might be thought of as the share of trend private stocks held for precautionary motives, or what FAO has termed the "reserve" component of stocks.

Substituting (4.1)-(4.3) in (2.8) an expression for trend excess supply can be obtained:

$$NX_t^* = (1 - (\delta + \varepsilon CV) \overline{Q}_t - \overline{D}_t + (d[1 - (\delta + \varepsilon CV)] + e) P_t^* \tag{4.4}$$
$$+ (\delta + \varepsilon CV) (\overline{Q_{t-1}} + dP_{t-1}) - (1 - \beta) \Delta SG_t^*.$$

Assume now that the trend domestic price is a constant multiple of the trend international price

$$P_t^* = \vartheta P_{wt}^* \tag{4.5}$$

where P_{wt}^* denotes the international (world) price trend. The coefficient ϑ-1 can be interpreted as an average rate of protection. If $\vartheta > 1$, then there is average protection, while if $\vartheta < 1$, there is average taxation. If a subscript i is added to each variable in (4.4) and (4.5) except the world price, and if the world equilibrium condition (which states that the sum of all excess supplies must equal zero) is considered, then the trend world price can be obtained:

$$P_{wt}^* = \sum_i \vartheta_i \left((d_i (1 - (\delta_i + \varepsilon_i CV_i)) + e_i\right)^{-1} \sum_i (\overline{D}_{it} - \tag{4.6}$$
$$- (1 - (\delta_i + \varepsilon_i CV)) Q_{it} - (\delta_i + \varepsilon_i CV)$$
$$(\overline{Q_{i,t-1}} + d_i \vartheta_i P_{w,t-1}^*) + (1 - \beta_i) \Delta SG_{it}^*.$$

It is not easy to predict trend world price from trend world demand and production. However, (4.6) clearly predicts that if there is a positive trend to publicly held stocks, then this has an unambiguous positive effect on average world prices, irrespective of what other protectionist or taxation policies are in place. It also predicts that a protective policy ($\vartheta > 1$) depresses mean world prices, while a taxation policy ($\vartheta < 1$) raises it.

Under the above model, the trend level of world private stocks will be:

$$SP_t^* = \sum_i SP_{it}^* = \sum_i [(\delta_i + \varepsilon_i CV_i) Q_{it}^* - \beta_i SG_{it}^*]. \tag{4.7}$$

A specification such as (4.7) shows that the two key components of trend private stocks are the degree of production fluctuations (expressed by the relevant coefficient of variation) and the trend of public stocks.

Price and Production Fluctuations

In this section, I outline a method for obtaining price and production fluctuations in the model outlined previously I shall first consider a closed economy and then extend it to a multiple country model.

Consider equation (2.9), set public stocks equal to zero, and set excess supply also equal to zero in order to simulate a closed economy. Then the following equation for deviation of price from trend is obtained:

$$(\alpha+a+c)p_t-\alpha p_{t-1}-\alpha p^e_{t+1/t}+(\alpha+b-c)p^e_{t/t-1}-x_t=0 \tag{5.1}$$

where for notational simplicity I have defined

$$x_t=u_t-v_t. \tag{5.2}$$

The solution of (5.1) will depend, of course, on the type of expectation formation that is assumed. Albeit the following results hold both for rational and adaptive expectations, I shall work with rational expectations, as they result in simpler expressions. The details of how to solve difference equations like (5.1) under rational expectations can be found in Muth (1961), Turnovsky (1979), and Sarris (1984) and will not be dealt with here. The upshot is that the expected price will obey the following equation:

$$p^e_{t+k/t}=\rho^k p_t \tag{5.3}$$

where ρ is the smallest root of the quadratic equation:

$$-\alpha\rho^2+(a+b+2\alpha)\rho-\alpha=0. \tag{5.4}$$

The actual price, in turn, obeys the following equation:

$$p_t=\rho p_{t-1}+\frac{x_t}{a+c+\alpha(1-\rho)}. \tag{5.5}$$

From (5.5), the asymptotic variation of price can be found (Sarris, 1984 or Turnovsky, 1979) as follows:

$$\sigma^2_p=\frac{\sigma^2}{(1-\rho^2)(a+c+\alpha[1-\rho])^2} \tag{5.6}$$

where σ^2 is the variance of x_t.

The parameter ρ, defined above, can be interpreted as a "decay" factor, in the sense that a shock in one year, that leads to a nonzero price deviation from trend will have effects that last for more than the current year but with a diminishing force. As an example, consider a negative random supply shock in one year t that leads to an unanticipated positive deviation of price from trend; namely, a positive value for p_t. This market effect implies that private agents will de-stock in this year, and, hence, will carry into the following year fewer than normal stocks. If production in year t is normal and there are no other random shocks, then private agents will be trying to build up stocks to normal in year t+1, and, hence, the total supply in year t+1, under normal and expected conditions in year t+1, will be smaller than what would occur had the shock in year t not occurred. Hence, price in year t+1 will be expected to be above normal, and this is the sense in which shocks in one year are expected to have aftereffects.

It can be shown (Sarris, 1984) that storage stabilizes prices and that the more responsive private agents are to expected price changes; namely, the larger is the value of parameter α, the smaller is the asymptotic price variance. It can also be shown that more elastic slopes of the supply and demand curves; namely, larger values of **a**, **b**, or **c** lead to smaller price variations. These properties are all quite reasonable and expected. Analytically, they can be written as follows:

$$\frac{\partial \sigma_p^2}{\partial \alpha} < 0 \tag{5.7a}$$

$$\frac{\partial \sigma_p^2}{\partial a} < 0 \tag{5.7b}$$

$$\frac{\partial \sigma_p^2}{\partial b} < 0 \tag{5.7c}$$

$$\frac{\partial \sigma_p^2}{\partial c} < 0. \tag{5.7d}$$

Suppose now that the public sector tries to stabilize the domestic prices by accumulating stocks when the price is below trend and by selling stocks when the price is above trend. A rule like that can be approximated as follows:

$$sg_t = -\zeta p_t \tag{5.8}$$

where $\zeta > \mathbf{0}$.

There are two approximations and simplifications inherent in (5.8). First, that public stocks rarely if ever hit zero. Otherwise, the rule becomes nonlinear and analytically intractable. Second, and this is more serious, that the public knows the underlying trend value of price $\mathbf{P}^*_t$. This is usually not the case and is bound to introduce a random element to (5.8) due to prediction inaccuracies. This term is neglected for the time being.

If a rule like (5.8) is incorporated in equation (2.10), then the form of solution under rational expectations does not change. The solution for the equilibrium price turns out to be the following:

$$p_t = \rho p_{t-1} + \frac{x_t}{a + c + \zeta(1-\gamma) + \alpha(1-\rho)} \tag{5.9}$$

where ρ now satisfies the following equation analogous to (5.4):

$$-\alpha\rho^2+(a+b+\zeta[1-\gamma]+2\alpha)\rho-(\alpha+\zeta[1-\gamma])=0. \tag{5.10}$$

Again, one of the solutions of (5.10) is between 0 and 1, and this is the stable solution of the model.

From (5.9) and (5.10), the asymptotic price variance can be shown to be equal to the following:

$$\sigma_p^2=\frac{\sigma^2}{(1-\rho^2)(a+c+\zeta[1-\gamma]+\alpha[1-\rho])^2}. \tag{5.11}$$

It can also be shown that

$$\frac{\vartheta\sigma_p^2}{\vartheta\zeta}<0 \tag{5.12}$$

or that stabilizing buffer stockholding by the government will indeed stabilize the market. If, of course, the public cannot forecast the trend accurately, then it might end up destabilizing the current price, depending on the forecast errors which would enter as an additional term in the numerator of (5.11).

Consider now a world composed of many countries each of which insulates its domestic market from the international one via a transmission coefficient η_i where

$$P_{it}=\eta_i(P_{wt}-P_{wt}^*)=\eta_i P_{wt}. \tag{5.13}$$

The analysis for many countries, each with a different price transmission coefficient η_i , is identical to the one exhibited above except that weighted averages of all the relevant parameters must be used. The result is that the world price must obey a relation of the following type

$$P_{wt}=\rho P_{w,t-1}+\frac{x_t}{a^*+c^*+\zeta^*+\alpha^*(1-\rho)} \tag{5.14}$$

where ρ satisfies the equation:

$$-\alpha^*\rho^2+(a^*+b^*+\zeta^*+2\alpha^*)\rho-(\alpha^*+\zeta^*)=0 \tag{5.15}$$

and where

$$x_t=\sum_i x_{it}=\sum_i (u_{it}-v_{it})=u_t-v_t \tag{5.16}$$

$$\zeta^*=\sum_i \eta_i\zeta_i(1-\gamma_i) \tag{5.17}$$

$$a^*=\sum_i \eta_i a_i \tag{5.18}$$

$$b^*=\sum_i \eta_i b_i \tag{5.19}$$

$$c^* = \sum_i \eta_i c_i. \tag{5.20}$$

The asymptotic world price variance under the assumption that individual country stocks are uncorrelated (for evidence of this, see Oskam and Blom, 1991) is given by the following expression:

$$\sigma_{pw}^2 = \frac{\sum_1 \sigma_i^2}{(1-\rho^2)(a^*+c^*+\zeta^*+\alpha^*[1-\rho])^2} \tag{5.21}$$

where

$$\sigma_i^2 = \sigma_{ui}^2 + \sigma_{vi}^2 - 2r_i \sigma_{ui} \sigma_{vi} \tag{5.22}$$

and where σ_{ui}, σ_{vi} denote the standard deviation of the random terms affecting domestic production and consumption, respectively, while r_i is the correlation coefficient between u_{it} and v_{it}.

The deviation of world production from trend can be found by summing the deviations of each country's individual productions from trend. The resulting expression after substitution is:

$$q_t = \Sigma q_{it} = \rho b^* P_{w,t-1} + \frac{c^* x_t}{(a^*+c^*+\zeta^*+\alpha^*[1-\rho])} + v_t. \tag{5.23}$$

The asymptotic variance of q_t can be found by first repeatedly applying (5.14) and then taking expectations of the squared expressions.

Given the above specifications, it is possible to write expressions for each country's production, stock, and trade deviations from trends only as functions of world price deviations from trends. These equations in elasticity form are as follows:

$$\frac{q_{it}}{Q_{it}^*} = \frac{\eta_i \mu_{is}}{\vartheta_i} \frac{P_{wt}}{P_{wt}^*} + \frac{\rho \eta_i}{\vartheta_i}(\mu_{il}-\mu_{is})\frac{P_{wt-1}}{P_{wt-1}^*} + \frac{v_{it}}{Q_{it}^*} \tag{5.24}$$

$$\frac{s_{it}}{Q_{it}^*} = -\frac{\eta_i}{\vartheta_i}(\tilde{\alpha}_i[1-\rho]+[1-\gamma_i]\tilde{\zeta}_i)\frac{P_{wt}}{P_{wt}^*} \tag{5.25}$$

$$\frac{nx_{it}}{Q_{it}^*} = \frac{\eta_i}{\vartheta_i}(\mu_{is}+\frac{v_i}{s_i} - [\tilde{\alpha}_i(1-\rho)+(1-\gamma_i)\tilde{\zeta}_i])\frac{P_{wt}}{P_{wt}^*} + \frac{\eta_i}{\vartheta_i}(\rho[\mu_i l-\mu_i s]+ \tag{5.26}$$

$$+ [\tilde{\alpha}_i(1-\rho) + (1-\gamma_i)\tilde{\zeta}_i]) \frac{P_{wt-1}}{P_{wt-1}^*} + \frac{v_{it}-u_{it}}{Q_{it}^*}$$

where

μ_{is}	=	shortrun (i.e., within a year) price elasticity of domestic production for country i
μ_{il}	=	longrun price elasticity of production for country i

a_i = elasticity of private speculative stocks (as a share of trend production) with respect to domestic price changes in country i

ς_i = elasticity of public stocks (as a share of trend production) with respect to domestic price changes in country i

Q_{it}^* = trend production in country i

P_{wt}^* = trend world price.

In (5.24), s_{it} represents the deviation of total (public and private) country i end-of-period stocks from their trend.

An interesting thing to notice from (5.24)-(5.26) is that the degree of price transmission η_i is a major influence on all variables. A very low value of η_i would render the deviations of domestic production, consumption, and stocks from their trends almost totally random. Another interesting observation is that lags play an important role in both domestic production and in excess supply.

Empirical Specification

Equations (5.15)-(5.17) allow one to assess the changes in the variability of production, consumption, stocks, and net exports from changes in overall trade policies summarized in the average protection parameters θ_i and the price transmission parameters η_i.

One of the major problems with models that attempt to project the levels of prices after liberalization is that the values of the various elasticities assumed are not internally consistent. A model that computes variabilities of prices and production imposes internal consistency on the various parameters.

This can be seen by computing the coefficients of variation of production and prices, which will be functions of the underlying parameters, and noting that, if the chosen elasticities are to be consistent they must result in price and production variabilities that are equal to the observed ones.

For the empirical results, time series annual data from 1970-89 on quantities grain produced, end-of-period total stocks, exports and imports of wheat, rice (in milled equivalent), and all coarse grains aggregated together were provided by FAO for two major aggregates of the world. The first consists of all developed market economies (DMC,s), while the other consists of all other countries; namely, the rest of the world (ROW). This latter group basically consists of all developing countries plus centrally planned economies. Furthermore, monthly data on international prices of wheat (No. 2, hard winter ordinary f.o.b. Gulf), yellow maize (U.S. No. 2, f.o.b. Gulf), and rice (Thailand, white 100 percent second grade, f.o.b., Bangkok) was provided for 1970-90. From this price series, average prices for crop years were obtained by averaging the July-June data.

After detrending, the regressions indicated in equations (5.15)-(5.17) were done. The results, however, did not produce robust estimates of the aggregated parameters. Hence, a different procedure was used. This procedure estimated directly the world price equation (5.14), and by utilizing estimated values for elasticities and the distortion parameters η and θ from other studies (mainly Tyers and Anderson, 1986; OECD, 1985, 1989a, 1989b; and, USDA, 1988), the remaining aggregate parameters were obtained. The stock parameters were obtained by estimating some equations directly and by calibrating the other parameters to fit observed variabilities. This procedure is lengthy and is described in detail in Sarris (1990). The results of the computations and calibrations are summarized in Table 2.1.

Impact of Trade Liberalization

Table 2.2 presents the effects of trade liberalization in wheat, rice, and coarse grains on the coefficients of variation of world production, and price of these products, and production within each block of countries.

For the wheat market, note that trade liberalization always lowers the coefficient of variation (CV) of world price. Also, the CV of world production goes down. However, the coefficients of variation of production of individual country blocs do not always decline and, in fact, they might increase in some cases. The explanation is that when a market is restricted and price transmission is relatively small, then even if there exists substantial shortrun price elasticity of supply, it will not cause production to vary by much since the price signals are not transmitted. When, however, price signals are transmitted, there is a lot more shortrun supply variability and it might destabilize production in a given region.

The instability of world and, especially, DMC rice production is also seen to increase under all trade liberalization scenarios, despite the fact that world price variability is decreased in all cases. This instability in DMC's largely caused by a very large short run price elasticity of supply in DMCs (2.057) which, under liberalization, causes production to vary substantially.

Although trade liberalization leads to declines in price variability in the coarse grain market, it also seems to lead to increases in production variability in some cases. The upshot of the results is that although trade liberalization seems to lead unambiguously to declines in world price variability, it could lead to small increases in world production variability. This would happen despite the fact that the short run price elasticities of supply are generally quite small.

As mentioned earlier, total stocks are composed of private and public stocks, and were distinguished as "trend" (normal end-of-season stocks) and speculative stocks. In Table 2.3, I show some basis statistics of total end-of-season stocks of cereals always measured as a ratio of trend production in the relevant region.

Table 2.1 Estimated coefficients for the model

Parameter		Wheat	Rice	Coarse grains
ρ		0.558	0.398	0.560
θ_i	DMC	1.190	2.490	1.110
	ROW	1.160	1.180	0.940
η_i	DMC	0.390	0.330	0.820
	ROW	0.460	0.500	0.640
μ_{ii}	DMC	0.550	0.330	0.450
	ROW	0.150	0.160	0.180
ν_i^*	DMC	0.090	0.100	0.200
	ROW	0.240	0.170	0.180
$\tilde{\alpha}_i$	DMC	0.813	1.829	0.359
	ROW	0.752	0.199	1.150
σ_i	DMC	0.400	0.400	0.400
	ROW	0.200	0.200	0.200
$\tilde{\zeta}_i$	DMC	0.406	0.915	0.180
	ROW	0.376	0.099	0.575

$*$ ν = absolute value of price elasticity of demand
DMC = Developed market economies
ROW = Rest of the world
Note: The values of the short run supply price elasticities were assumed at half their long run values indicated above.

It can be seen that world stocks in individual cereals have been increasing slowly, as a share of trend production, over the last two decades for each individual product.

Total cereal stocks average 19.4 percent of world production in 1971-89. For the four years 1985-89 they averaged 19.1 percent, implying a relative constancy of total end-of-season world cereal stocks as a share of total world cereal production. Notice that in 1985-89 the share of wheat and rice stocks in production in both DMC's and the ROW have been below their long-term average levels, while they have been above normal for course grains, especially in DMCs. However, the share has stayed roughly constant for all cereals.

There are no data for the portion of world stocks that is held by public or private sources. For large country aggregates this might not be an appropriate distinction, because, in some large cereal producing countries, notably the USSR, all stocks are held by public authorities and there is probably very little trade-off between public and private stocks. For the few countries for which data were available for 1979-89, there is no clear pattern of growth or decline in private or public stocks over the last decade. However, there are some countries (all in the group classified earlier as ROW) that exhibit nonzero average levels of public stocks as a share of trend production. One could infer

Table 2.2 Impact of trade liberalization for wheat, rice, and coarse grains on variability of world production and prices (all coefficients of variation are expressed in percentages)

	Before trade liberalization	Free trade		
		DMC only	in ROW only	DMC and ROW
Wheat				
CV_{qw}	4.300	4.150	3.544	3.732
CV_{pw}	24.700	18.250	13.260	4.110
CV_{q1}	7.900	8.920	7.830	7.971
CV_{q2}	5.300	5.660	4.290	4.166
Rice				
CV_{qw}	2.800	3.050	2.780	2.860
CV_{pw}	36.800	24.260	19.200	14.440
CV_{q1}	6.300	44.850	8.180	28.420
CV_{q2}	3.000	2.930	2.890	2.890
Coarse grains				
CV_{qw}	5.700	5.750	5.640	5.690
CV_{pw}	23.900	20.710	22.430	20.170
CV_{q1}	11.340	11.160	10.650	10.850
CV_{q2}	4.560	4.090	3.990	4.950

Legend: CV_{qw} — Coefficient of variation (CV) of world production
  CV_{pw} — CV of world price
  CV_{q1} — CV of production in DMC
  CV_{q2} — CV of production in ROW

from this admittedly very limited evidence that there is a significant part of ROW trend cereal stocks that is handled by the public sector, while for DMC's the share of public stocks in "trend" or normal total stocks is quite small, although there might be large amounts of occasional "speculative" stocks.

In the sequel, I shall assume that the share of trend public cereal stocks in total trend stocks (always as a share of trend production) is 0.5 for the ROW and 0.1 for the DMC's. Denote this share by φ_{is}. Using earlier notation:

$$\varphi_{is} = \frac{SG_i^*}{SP_i^* + SG^*} = \frac{SG_i^*/Q_i^*}{(SP_i^* + SG_i^*)/Q_i^*}. \tag{7.1}$$

Then, for a country i, equation (4.7) can be written as:

$$ST_i^*/Q_i^*(1 - [1 - \beta_i]\varphi_{is}) = \delta_i + \varepsilon_i CV_{qi} \tag{7.2}$$

where ST_i^* denotes the total (private plus public) end-of-season stocks.

In equation (7.2), we have data on the average share of total stocks in trend production and we have estimated the coefficients of variation of production. However, there are still three parameters that must be specified; namely, β_1, δ_1,

Table 2.3 Statistics of end of season cereal stocks as a share of trend production (data for 1971-1988)

	Mean	Standard deviation	Coefficient of linear trend[1]	Average 1985-1989[2]
Wheat				
DMC	0.359	0.076	0.004	0.318
ROW	0.190	0.035	0.004**	0.177
World	0.252	0.041	0.004**	0.226
Rice				
DMC	0.275	0.121	-0.002	0.178
ROW	0.150	0.019	0.002**	0.123
World	0.157	0.020	0.002**	0.126
Coarse Grains				
DMC	0.253	0.112	0.015**	0.320
ROW	0.102	0.012	0.001**	0.090
World	0.172	0.054	0.007**	0.195
All Grains	0.191			

Source: Computed from FAO data.

[1] An asterisk means significance at 90 percent level. Two asterisks mean significance at 95 percent level.

[2] For this column average 1985-1989 end of season stocks have been divided by average 1985-1989 production (FAO data).

and ϵ_i. If two of these are specified, then equation (7.2) could be used to compute the third. For the sequel, the following values will be assumed. For β_i, namely, the trade-off coefficient between trend private and public stocks (as contrasted with the coefficient γ_i for trade,off between deviations from trend private and public stocks), a value of 0.5 will be assumed for DMC's and a value of 0.3 will be assumed for ROW. The justification for this choice is that when the private sector has built up expectations about the constancy of the public stock level, then private agents will be more likely to de-stock than when only a temporary public stock accumulation takes place. Hence, a value of β_i larger than γ_i is in order.

The crucial parameter is ϵ_i. It denotes the response of private end of season stocks to the equilibrium variability of production. Consider Figure 2.2. That figure was used to argue that when the distribution function of production becomes flatter (namely, when the production variance becomes larger), then the precautionary stock level also becomes larger. For lack of empirical data, it will be hypothesized that the marginal trend private stock level responds with unity coefficient to variations in the coefficient of variation of production. In other words, a value of ϵ_i equal to one will be assumed. Using equation (7.2), then the values for δ_i are implied, using the estimated values for CV_{qi} from Table 2.2 and the average shares of total stocks in trend production for the period 1985-89 reported in Table 2.3. The recent period is used because trade liberalization must be examined relative to the current situation.

Although the mean level of deviations of stocks from trends analyzed earlier is zero, it will be of interest to analyze the"speculative element" (the variability of this deviation), which equation (5.17) shows is related to the coefficient of variation of world price, the relevant stock adjustment parameters and the "equivalent transmission coefficient" η_i/θ_i. Notice that although the coefficient of variation of price will decline under trade liberalization, the coefficients η_i/ϑ_i will increase hence leading to an indeterminate change for the mean speculative stock level. However, for the world as an aggregate, it can be shown (Sarris, 1990) that the standard deviation of speculative stocks

SR_w^* as a fraction of world trend production is equal to:

$$\frac{SR_w^*}{Q_w^*}=(\frac{CV_u^2+CV_v^2}{1-\rho^2})^{1/2}$$

(7.3)

where CV_u and CV_v are the coefficients of variation of the world random production and demand stocks respectively. Hence, the change in the mean speculative stock level is determined by the parameter ρ.

Table 2.4 summarizes the results of trade liberalization. The general pattern that emerges is that "trend" or "normal" world carryover stocks decrease as a share of trend production under all liberalization scenarios, although this sometimes hides increases by one group of countries and decreases by another. There are two reasons. The first is that public trend stocks decline and, hence, the private sector increases its own normal stocks, but the increase is not enough to counteract this decline. The second is that production variability in many cases decreases. The average level--namely, the standard deviation of "speculative" stocks--is also seen to decline. This is mainly due to the general decline in world cereal price variability.

The trend level of world total cereal stocks as a proportion of trend production is seen to stay roughly unchanged at 19 percent when only DMC's liberalize, but declines to 15.7 percent when all countries liberalize. Given that an interpretation of δ_i could be made as the portion of total private trend stocks held for transactions purposes only (the FAO minimum working stock level),one can compute this under a complete trade liberalization scenario and it is 9.4 percent. Hence, the difference between 15.7 percent and 9.4 percent or 6.3 percent could be termed the longrun precautionary or "reserve" stock level. These figures, of course, depend on the particular values of the parameters assumed and must be examined further by appropriate sensitivity analysis.

Table 2.4 Effects of trade liberalization on cereal stock levels (expressed as proportions of regional trend production)[1]

Commodity	Item	Before trade liberalization	Free trade DMC only	ROW only	DMC and ROW
Wheat					
DMC	ST*/Q*	0.318	0.312	0.317	0.303
	SR*/Q*	0.049	0.075	0.029	0.002
ROW	ST*/Q*	0.177	0.182	0.105	0.104
	SR*/Q*	0.063	0.050	0.052	0.002
World	ST*/Q*	0.226	0.228	0.180	0.174
	SR*/Q*	0.058	0.059	0.044	0.002
Rice					
DMC	ST*/Q*	0.178	0.555	0.198	0.390
	SR*/Q*	0.073	0.243	0.042	0.172
ROW	ST*/Q*	0.123	0.122	0.079	0.079
	SR*/Q*	0.031	0.019	0.026	0.019
World	ST*/Q*	0.126	0.142	0.084	0.094
	SR*/Q*	0.033	0.030	0.027	0.026
Coarse Grains					
DMC	ST*/Q*	0.320	0.307	0.312	0.300
	SR*/Q*	0.047	0.041	0.048	0.039
ROW	ST*/Q*	0.090	0.083	0.053	0.065
	SR*/Q*	0.157	0.153	0.129	0.125
World	ST*/Q*	0.195	0.185	0.171	0.171
	SR*/Q*	0.107	0.100	0.092	0.086
All Cereals					
DMC	ST*/Q*	0.316	0.315	0.311	0.303
	SR*/Q*	0.048	0.057	0.042	0.031
ROW	ST*/Q*	0.126	0.124	0.076	0.080
	SR*/Q*	0.093	0.084	0.076	0.057
World	ST*/Q*	0.191	0.190	0.157	0.157
	SR*/Q*	0.078	0.074	0.064	0.034
TOTAL[2]		0.269	0.264	0.221	0.191

Source: Computed.

[1] ST*/Q* denotes the "trend" or normal level of carryover stocks. SR*/Q* denotes the value of the standard deviation of the reserve level of stocks as a proportion of trend production.

[2] This number is only indicative as it does not represent any particular level.

There are two implications of these forecasts for food security in developing countries. First, because there will be less stock available for concessionary emergency assistance under trade liberalization the level of publicly held stocks declines. There will also be fewer total private stocks available, and these will

Table 2.5 Impact on the coefficient of variation of consumption of ROW from trade liberalization (in percentages)

	Wheat	Rice	Coarse grains
CV_{d2} under			
Current situation	6.50	4.37	15.78
Free trade in DMC	6.75	5.00	15.95
Free trade by all	7.20	4.41	15.25

Source: Computed.

have to be procured commercially in times of shortage.

Second, world price variability will decline and this creates less of a need for large levels of emergency stocks held against large international price rises. The probability of a price spike likely decreases as the overall price variability declines, but it is quite possible that the magnitude of the price spike might be very large.

Concerning the level of world stocks under the most likely trade liberalization scenario, which is liberalization only by DMC's, there are two counteracting effects. First, the normal end-of-season private stocks as a share of world trend production would show marginal decline (Table 2.4). The decrease in "speculative" stocks from 7.8 percent to 3.4 percent, in turn, implies that the world cereal stock fluctuations about the slightly lower trend would be smaller.

Table 2.5 indicates the computed coefficient of variation of the consumption in the ROW. The table reveals that in all cases, trade liberalization in cereals by DMC's will destabilize consumption in the ROW, if the ROW does not liberalize. However, if the ROW also liberalizes, the situation improves in only the coarse grain market, and it is still worse in the wheat market. The upshot is that trade liberalization will most likely destabilize consumption in the ROW, and probably in the LDC's that are part of the ROW.

The reasoning behind this somewhat unexpected result lies in the fact that the coefficient of variation of purely random consumption shocks CV_{ui} is large. The variations in domestic consumption,are first due to the variations in world price, which, when transmitted to the domestic market, influence current consumption along the domestic demand curve. Secondly, and more importantly, they are due to the random shocks in demand, whose influence, however, is tempered by the degree to which the domestic consumption shocks influence world price. When world price variability is large, it tends to substantially offset the domestic demand shocks. However, under trade liberalization, world price variability declines, and thus domestic consumption is more susceptible to the domestic random shocks, which are large.

Conclusion

The analysis in this chapter shows that agricultural trade liberalization has complex implications for world price, production, and stock variabilities. Stock adjustment is quite substantial in the current world cereal markets and would play an even more significant role in stabilizing world prices after trade liberalization. However, trend levels of world stocks are likely to fall after liberalization, as the public sector would lessen its interference in cereal markets and the private sector most likely would not fully compensate for the reduction.

World price fluctuations were seen to be substantially lessened after cereal trade liberalization. This is outcome to be expected, as the majority of countries tend to adopt policies that stabilize domestic prices at the cost of destabilizing international ones. Given the inelasticity of world supply and demand curves, this effect was shown to be quite strong.

Developed market economies as a whole tend to insulate their domestic cereal markets more than the developing and centrally planned economies aggregated together, and that price transmission elasticities seem to be smaller for wheat and rice as compared to coarse grains. The former effect might be justified on the grounds that DMCs have more resources as well as better administrative control of their economies. The latter could be explained by the fact that the United States, which is a relatively open economy as far as coarse grains trade is concerned, has an overwhelming share of world coarse grain trade and has a very large share of DMC's production.

Despite the substantial reduction in world price variability predicted under cereal trade liberalization, production is not greatly stabilized, and for some products and some regions it is even somewhat destabilized. This lack of stability is the result of short run production response that now operates on relatively more variable domestic prices, because of increases in price transmission.

Transactions and precautionary stock levels, analogous to FAO minimum working stocks and reserve stocks were modeled separately and in simple fashion. Public crowding out of private stocks was also explicitly incorporated. Under these conditions, changes in price and production variabilities have direct implications for world cereal stock levels. It was seen that average world cereal stock levels as a share of world production do not significantly change if DMC's only liberalize, and will fall somewhat if all countries liberalize. In fact, under the parameter assumptions made, the world cereal stock levels after all countries adopted complete trade liberalization might fall to levels below 17-18 percent of world production. The 18 percent level has been previously estimated by FAO (1983) as "safe" for world food security. Also speculative stock variations in turn, which are short term responses to current unexpected shocks, will tend to be smaller under trade liberalization because world prices will be less variable.

References

Blanchard, O. J., and S. Fischer. 1989. *Lectures on Macroeconomics*. Cambridge, Massachusetts: MIT Press.

Burniaux, J. M., F. Delorme, I. Lienert, and J. Martin. 1990. *WALRAS--A Multi-Sector Multi-Country Applied General Equilibrium Model for Quantifying the Economy-Wide Effects of Agricultural Policies*. OECD Economic Studies, No. 13, Winter 1989-90.

Evans, M. K. 1969. *Macroeconomic Activity: Theory Forecasting and Control: An Econometric Approach*. New York: Harper and Row.FAO. 1983. *Approaches to World Food Security*. Economic and Social Development Paper No. 32, Rome.

Ghosh, S., C. L. Gilbert, and A. J. Hughes Hallett. 1987. *Stabilizing Speculative Commodity Markets*. Oxford: Clarendon Press. Kaldor, N. 1939. "Speculation and Economic Stability." *Review of Economic Studies*, Vol. 7, 1-27.

Muth, J. F. 1961. "Rational Expectations and the Theory of Price Movements." *Econometrica,.* Vol. 29, 315-335.

Newbery, D. M. G., and J.E. Stiglitz. 1981. *The Theory of Commodity Price Stabilization*. Oxford: Clarendon Press.

OECD. 1985. Model Specification and Data, Joint Working Party of the Committee for Agriculture and the Trade Committee, Working Party No. 2 (Commodity Analysis and Market Outlook). DAA/1925 - TD/85.261, September 17.

OECD. 1989a. PSE and CSE Tables, 1979-1988 Paris. October 1989.

OECD. 1989b. *Agricultural Policies in Developing Countries and Agricultural Trade: Statistical Annex*. Committee for Agriculture, Trade Committee, AGR98(2), TC(89)4, February 16.

Oskam, A., and J. Blom. 1991. *Instability in Grain Production and Trade*. Mimeographed. The Netherlands: Department of Agricultural Economics, Wageningen Agricultural University, April.

Parih, K. S., G. Fischer, K. Frohberg and O. Gulbrandsen, 1988.*Towards Free Trade in Agriculture*. Laxenburg, Austria: Martinus Nijhoff Publishers.

Roningen, V. O., and P. M. Dixit. 1989. *Economic Implications of Agricultural Policy Reforms in Industrial Market Economics*. U.S. Department of Agriculture, Economic Research Service, Staff Report No. AGES 89-36, August.

Sarris, A. H. 1984. "Speculative Storage, Futures Markets and the Stability of Agricultural Prices." Chapter 3 in Storey, G. G., A. Schmitz, and A. H. Sarris, *International Agricultural Trade: Advanced Readings in Price Formation, Market Structure, and Price Instability*. Boulder, Colorado: Westview Press.

Sarris, A. 1990. *The Impact of Agricultural Trade Liberalization on World Cereal Instability and Stocks.* Paper prepared for FAO, Commodities Division, May. Turnovsky, S. J. 1979. "Futures Markets, Private Storage, and Price Stabilization." *Journal of Public Economics,* Vol. 12, 301-327.

Tyers, R., and K. Anderson. 1986. *Distortions in World Food Markets: A Quantitative Assessment.* Background paper prepared for the World Bank 1986. World Development Report, January.

USDA. 1988. *Agriculture in the Uruguay Round: Analyses of Government Support.* Staff Report No. AGES880802. Economic Research Service, Agriculture and Trade Analysis Division, December.

Valdés, A., and J. Zietz 1980. *Agricultural Protection in OECD Countries: Its Cost to Less-Developed Countries.* International Food Policy Research Institute, Research Report 21, Washington, D.C.

Weymar, F. H. 1968. *The Dynamics of the World Cocoa Market.* Cambridge, Massachusetts: MIT Press.

3

Implications of Grain Trade Liberalization for LDC Food Security

Peter Hazell

Introduction

The Uruguay Round of the General Agreement on Trade and Tariffs (GATT) trade negotiations has stimulated numerous studies of the possible economic effects of trade liberalization on world grain markets. A recent Organization for Economic Development (OECD) and World Bank study (Goldin and Knudsen, 1990) has tried to make some sense of the often conflicting results obtained from these studies. Their general conclusion is that average world grain prices are likely to increase, but only modestly, particularly if both the OECD and developing countries liberalize their domestic grain markets. The variability of world grain prices also seems likely to decline as increased trading opportunities serve to pool production risks more globally. Anderson and Tyers (1990) claim that the coefficient of variation (cv) of world grain prices could decline by as much as one-third.

These changes would likely be beneficial to the developing countries (LDC's) as a group, particularly if liberalization also meant removal of restrictions on OECD imports of tropical agricultural products and their processed derivatives. LDC's would face more stable prices, and domestic agricultural growth would become more economically viable. This, in turn, would increase incomes and employment in rural areas, adding to the food purchasing power of poor people.

Potential Problems for Developing Countries

There are two potential difficulties with the trade liberalization scenario that could impact negatively on the food security of the poorest countries. First, liberalization in OECD countries would reduce food surpluses, and this, in turn, is likely to reduce food aid. For several of the poorest LDC's, especially in sub-Saharan Africa, this could lead to a significant worsening of their trade balances. The problem would, of course, be aggravated by any increase in average grain prices. An obvious solution is for OECD countries to increase alternative forms of bilateral aid.

Second, publicly held food stocks would decline under trade liberalization, raising the possibility of significant price spikes in years when global production falls significantly below trend. In 1988/89, for example, world total grain production was 114 million tons (or 6.8 percent) below trend and, in the absence of sizeable global stocks that year (about 400 million tons), prices could have increased markedly. Price spikes need not be inconsistent with the predicted decline in overall price variability. They simply correspond to unusual events in the upper tail of the price distribution or to "catastrophic" events in insurance parlance.

The risk of global production shortfalls below trend is also increasing over time because of increasing variability in world grain production. As shown in Table 3.1, the cv of world grain production around trend has increased since the 1960's, and the probability that world production falls 5 percent or more below trend has doubled (from about 0.04 to 0.08). As I have shown elsewhere (Hazell, 1985), this increase in production variability is largely attributable to more variable yields that have also become more positively correlated between regions and crops.

Increasingly, yield variability is behaviorally induced, because more farmers are adopting input-responsive varieties and are becoming better integrated into the global economy. As world prices fluctuate, increasing numbers of farmers are not only exposed to the same price signals (and this would increase further with trade liberalization), but they also adjust their input use in the same direction and at the same time. This increasing synchronization of yield variations is particularly pronounced between regions within countries, as I have shown for India and the United States (Hazell, 1984). The pattern may be aggravated by the fact that large crop areas are now planted to the same or closely related varieties, which have a common susceptibility to the same pest, disease, and weather stresses. The price spike problem should not be overstated. During the 28-year period 1963/64 to 1990/91, world grain production only fell 4 percent or more below trend in 3 of those 28 years.

Even in the absence of sizeable government-owned stocks, private storage, together with the potential buffering effects of the livestock feedgrain market, ought to be able to moderate the price effects of most global production

Table 3.1 Variability of world grain production around a quadratic trend: 1963/64 - 1990/91

Period	Average production	Standard deviation		Probability of a 5 percent shortfall below trend
		Coefficient	Probability	
	(1000 mt)	(1000 mt)	percent	percent
Decade beginning:				
1963/64	1033	24.5	2.37	1.74
1964/65	1073	26.9	2.51	2.33
1965/66	1103	30.1	2.73	3.36
1966/67	1136	33.0	2.90	4.27
1967/68	1172	33.3	2.84	3.92
1968/69	1203	33.3	2.77	3.51
1969/70	1242	40.7	3.28	6.30
1970/71	1278	40.6	3.18	5.82
1971/72	1313	39.6	3.02	4.85
1972/73	1343	38.5	2.87	4.01
1973/74	1382	37.3	2.70	3.22
1974/75	1404	39.7	2.83	3.84
1975/76	1447	43.6	3.01	4.85
1976/77	1488	42.0	2.82	3.84
1977/78	1520	43.4	2.85	4.01
1978/79	1547	44.8	2.90	4.18
1978/80	1556	56.4	3.62	8.38
1980/81	1582	56.6	3.58	8.08
1981/82	1614	57.8	3.58	8.08

shortfalls. The problem could be greater for some individual cereals, particularly rice and white maize, since their world markets are relatively thin and more prone to production-induced price shocks.

International Solutions

There are two ways in which the international community can help LDC's cope with occasional price spikes. The first approach is for OECD countries to retain sufficient public grain stocks to cushion any major shortfall in global production. This possibility was analyzed quite extensively in the 1970's following the world food crisis of 1972/73. For example, Reutlinger (1976) and Sarris and Taylor (1976) undertook simulation studies and concluded that a global grain stock of 20 to 30 million tons would significantly reduce the risk of a major price spike. Such a stock was equivalent to nearly two standard deviations of world grain production at that time. An equivalent risk-reducing stock would have to be about 50 to 60 million tons today.

To be cost effective, an international grain stock should be managed purely as a price insurance mechanism. That is, stocks would only be released if a trigger price were exceeded, and purchasing rules would be directed at maintaining only the required stock. Unfortunately, such a stock management system is unlikely to benefit the grain exporting countries, as shown in both the Reutlinger and Sarris and Taylor studies. This lack of benefit to exporters complicates the possibility for attaining the required international agreement.

An alternative, and probably much more cost-effective solution, is for the OECD countries to offer foreign exchange insurance to LDC's to enable them to compete for food imports in years when prices are high (Huddleston et al., 1984). The International Monetary Fund (IMF's) Compensatory Finance Facility already plays this type of role in providing balance-of-payments support to compensate for precipitous increases in the cost of cereal imports. Export credits issued by some OECD countries also play a similar role. These kinds of arrangements might become more important than in the past should trade liberalization succeed.

National Solutions

At the national level, there are a number of things that governments can do to reduce their country's exposure to catastrophic price risks. In reality, these alternatives appear more promising than international solutions, and we should not lose sight of them in focusing the discussion of this working group on international stocks.

Many LDC's already seek to enhance their food security through programs of national self-sufficiency in basic food grains. This strategy can be costly if a country's comparative advantage lies with export crops. But some diversification into food crops and away from pure exploitation of comparative advantage is perfectly rational for a country facing volatile world prices and imperfect insurance markets as Jabara and Thompson (1980) and Sarris (1985) have shown.

Diversification for food security should go beyond basic grains. Many countries can also grow a variety of tubers, root crops, pulses, and plantains that have been relatively neglected in past agricultural research and development programs. Some of these crops are quite robust, particularly during droughts, and their yields can usefully offset losses in cereal production.

LDC's can also carry emergency food stocks of their own to cope with high prices in years of short supply years. Past attempts to use stocks to stabilize domestic grain prices have proved expensive and relatively ineffective in most countries (Knudsen and Nash, 1990). Stocks should only be carried as an insurance mechanism, and should be small in relation to national consumption. McIntire (1981) suggests a figure of 5 percent of national consumption, but this was for land-locked Sahelian countries. Even smaller stocks would be appropriate for most LDC's with easier access to international markets (World Bank, 1986).

Developing countries that depend on food imports could also protect themselves against price spikes by hedging in world futures and options markets. Surprisingly, few LDC's seem to do this, despite the fact that food imports are typically controlled by government-owned parastatals. At a time when LDC governments are being urged to phase out parastatals and to privatize grain marketing, it is hardly appropriate to suggest a new role for parastatals. But there is no reason why a ministry of finance, or a central bank, should not engage in futures trading in the national interest. The World Bank is currently offering technical advice to countries with this objective in mind. And, the bank has already helped to set up risk management units in several countries.

Developing countries can also self-insure by carrying a reserve of foreign exchange specifically earmarked for food imports. Such a reserve might be associated with a variable levy on food imports. The reserve would be built up in low-price years and drawn down in high-price years (for example, see Siamwalla, 1986).

In countries where the proportion of food-insecure households is relatively small, it may be more efficient to focus on targeted assistance programs (e.g., food subsidies) than on food security policies that distort domestic prices for all food market participants (Sahn and von Braun, 1987).

Finally, there is considerable scope in many developing countries for reducing grain production variability. Irrigation investments often act to reduce production instability while also increasing average productivity (Mehra 1981). Plant breeding and agronomic research can contribute to more stable yields, as can more reliable supplies of fertilizers, water, and electricity in rural areas (Anderson and Hazell, 1989). Interregional correlations in production can also be exploited to reduce aggregate production variability by focusing producer incentives and public investments to increase production in regions with low production variability. Investments can also be directed to regions in which production is negatively or only weakly correlated with the production of other important regions (Hazell, 1982: Tarrant, 1988). Lastly, improvement of infrastructure and marketing networks can facilitate greater inter-regional trade within countries, and this can have important risk pooling effects.

References

Anderson, Jock, and Peter Hazell. 1989. *Variability in Grain Yields: Implications for Agricultural Research and Policy in Developing Countries.* Baltimore: Johns Hopkins University.

Anderson, Kim and Rod Tyers. 1990. "How Developing Countries Could Gain from Agricultural Trade Liberalization in the Uruguay Round," in *Agricultural Trade Liberalization*, I. Goldin & O. Knudsen (eds). Paris: OECD and the World Bank.

Goldin, Ian, and Odin Knudsen (eds). 1990. *Agricultural Trade Liberalization: Implications for Developing Countries.* Paris: OECD and the World Bank.

Hazell, Peter. 1982. *Instability in Indian Foodgrain Production.* IFPRI Research Report No. 30. Washington, D.C.: International Food Policy Research Institute.

_____. 1984. "Sources of Increased Variability in Indian and US Cereal Production," *American Journal of Agricultural Economics*, 66(3): 302-11.

_____. 1985. "Sources of Increased Variability in World Cereal Production Since the 1960's," *Journal of Agricultural Economics*, 36(2): 145-59.

Huddleston, Barbara; D. Gale Johnson; Shlomo Reutlinger; and Alberto Valdés. 1984. *International Finance for Food Security.* Baltimore: Johns Hopkins University Press.

Jabara, Cathy and Robert Thompson. 1980. "Agricultural Comparative Advantage Under International Price Uncertainty: The Case of Senegal," *American Journal of Agricultural Economics*, 62(2): 188-98.

Knudsen, Odin and John Nash. 1990. "Domestic Price Stabilization Schemes in Developing Countries," *Economic Development & Cultural Change*, 38: 539-58.

McIntire, John. 1981. *Food Security in the Sahel: Variable Import Levy, Grain Reserves, and Foreign Exchange Assistance.* IFPRI, Research Report 26, International Food Policy Research Institute. Washington, D.C.

Mehra, S. 1981. *Instability in Indian Agriculture in the Context of the New Technology.* IFPRI Research Report No. 25. International Food Policy Research Institute. Washington, D.C.

Reutlinger, Shlomo. 1976. "A Simulation Model for Evaluating Worldwide Buffer Stocks of Wheat," *American Journal of Agricultural Economics*, 58(1): 1-12.

Sahn, David and Joachim von Braun. 1987. "The Relationship between Food Production and Consumption Variability: Policy Implications for Developing Countries," *Journal of Agricultural Economics*, 38(2): 315-28.

Sarris, Alexander and Lance Taylor. 1976. "Cereal Stocks, Food Aid and Food Security for the Poor," *World Development*, 4(12): 967-76.

Sarris, Alexander. 1985. "Food Security and Agricultural Production Strategies Under Risk in Egypt," *Journal of Development Economics*, 19: 85-111.

Siamwalla, Amar. 1986. "Approaches to Price Insurance for Farmers," in *Crop Insurance for Agricultural Development*, Peter Hazell, Carlos Pomareda and Alberto Valdés (eds). Baltimore: Johns Hopkins University Press.

Tarrant, John. 1989. "An Analysis of Variability in Soviet Grain Production," in *Variability in Grain Yields*, Jock Anderson and Peter Hazell (eds). Baltimore: Johns Hopkins University Press.

World Bank. 1986. *Poverty and Hunger: Issues and Options for Food Security in Developing Countries*. Washington, D.C.

4

Australia's Experience with
Its Wool Buffer Stock Scheme[1]

H. Don B.H. Gunasekera
and Brian S. Fisher

Introduction

Governments and sometimes private marketing agencies adopt various measures to maintain agricultural price stability. These measures usually involve some reduction in the level of prices during a boom and the supplementation of prices during a recession. This may be done, for example, by setting up a buffer stock scheme whereby an agency will buy some of the commodity when the market price falls below an agreed level (a floor price), with the object of raising the market price. The commodity is stored by the agency, and re-offered for sale when the market is more buoyant (Campbell and Fisher 1991).

Buffer stock schemes attempt to reduce price variability by transferring actual commodity from one period of sale to another. Whether there are net benefits to producers from such schemes is an empirical question that must be assessed on a case by case basis. Any buffer stock scheme must be viewed as a speculative operation and the judgements made about buying and selling are vital to the outcome. The fallibility of the management is without doubt the greatest hazard associated with any such scheme (Campbell and Fisher 1991). For example, stocks may be purchased rapidly by an agency when prices turn

---

[1] Published with permission of Blackwell Publishers, Oxford, UK. Previously published in *The World Economy*, Vol. 15.

downward after a boom. Although this may help the agency to maintain the floor price well above the level that would have been determined by market forces, a significant danger of this approach is the accumulation of large stock levels. Such was the experience of the (now defunct) Australian Wool Corporation during the latter part of the 1980s, which led to the collapse of its wool buffer stock scheme.

Those who are contemplating the establishment of a buffer stock scheme should seriously ask whether the potential gains are commensurate with the costs. Also, they need to consider the implications for the economy at large. In addition to the above concerns, one needs to consider the long term viability of such a scheme. The viability of a scheme will be critically determined by the stock disposal strategy adopted (Campbell and Fisher 1991). The stock disposal strategy is also critical even after the collapse of the scheme. This has been the experience in relation to the collapsed wool buffer stock scheme in Australia.

The purpose of this chapter is to use Australia's recent experience with the collapse of its wool buffer stock scheme in order to illustrate the problems involved (a) in using a buffer stock scheme and (b) in disposing of accumulated stocks, particularly after the collapse of such a scheme.

Background

Wool is one of the most important export industries in Australia. In recent years wool has accounted for nearly 30 per cent of Australian agricultural export earnings and around 10 per cent of total export earnings. Given wool's position as a key Australian rural industry, the economic efficiency of the Australian wool industry and the economic welfare of wool growers are of considerable national importance. Australia accounts for nearly one-third of the world's wool production, followed by the Soviet Union (15 per cent) and New Zealand (10 per cent). China, Argentina, South Africa, and Uruguay each account for 3-7 per cent of world production of wool. Around 80 per cent of the wool produced in Australia is exported as raw wool. The European Community is the largest market for Australian wool, taking, on average in the past decade, around one-third of Australian raw wool exports. Japan is the second largest buyer of Australian wool and is the largest individual country buyer. Japan accounts for around 19 per cent of Australian wool exports. In recent years, the Soviet Union, China, and Eastern Europe have emerged as important buyers of Australian wool, purchasing an average of 10 per cent, 8 per cent, and 6 per cent of Australia's wool exports, respectively.

The level of wool supplies in Australia is influenced by many factors including seasonal conditions, production costs, and returns from wool in both absolute terms and relative to returns from other agricultural commodities such as grains, beef, and sheep meat, all of which compete for farm resources. In addition, the level of stocks has a direct bearing on the availability of wool. Wool growers have only limited scope to adjust their production to changing market conditions in the short term because of the lead times required to change flock sizes and production regimes. Hence, accumulated stocks can play a key role in moderating short term production and demand imbalances. In the medium to longer term, however, wool production is more responsive to price changes. (For estimates supply elasticities see, Longmire, Brideoake, Blanks, and Hall, 1979: Fisher and Munro, 1983: and Fisher and Wall, 1990).

The factors that influence the demand for wool products, and, hence, for wool, are diverse and can be categorized into four broad groups:

- the price of wool and its relativity to the prices of other fibers, including synthetics and cotton,
- the availability of foreign exchange and the domestic demand for raw wool in key importing countries such as the Soviet Union and China,
- world economic activity and macroeconomic developments outside the fibre market, and
- and consumer preferences which are influenced by changes in fashion and by promotional initiatives.

In the recent past, the Australian wool industry has undergone significant changes, particularly with regard to wool pricing and marketing arrangements. The history of wool marketing arrangements in Australia dates back to the early 1970's. Following a period of rapidly falling wool prices, a scheme was set up in November 1970 to smooth price fluctuations rather than to defend a minimum price. Under this scheme' the then Australian Wool Commission began buying wool during the 1970-71 season, which was subsequently sold at a profit. The scheme was initially funded through a government established credit facility where borrowings were charged at a concessional rate of interest. Under this scheme, the Commission supported prices, but there was no publicly announced minimum price to be maintained. A number of changes were subsequently made to the scheme. The most notable of these changes were the establishment of the Australian Wool Corporation (AWC) and its authority to operate a flexible reserve price scheme for wool under the *Wool Industry Act* of 1972; the introduction of the producer funded Minimum Reserve Price Scheme in 1974; and the passage of the *1987 Wool Marketing Act*. The 1987 act devolved responsibilities for setting floor prices and the tax rate (which was used to partially fund the buffer stock scheme) to both the Australian Wool Corporation and the Wool Council of Australia (a body of grower elected representatives). The Minimum Reserve Price scheme was suspended in February 1991 and officially discontinued for the 1991-92 wool selling season and after.

Reserve Price Scheme

The Reserve Price Scheme was a type of buffer stock scheme in which an administering authority bought and sold wool in the marketplace, artificially adjusting wool demand and supply in an attempt to stabilise prices. However, the scheme differed from the classical buffer stock scheme insofar as there was no explicit attempt to maintain market prices within some predetermined band. Rather, the main emphasis appeared to have been to keep the market average price above an annually set minimum (the floor price). Clearly, however, the sale of stockpiled wool depressed market prices during those periods in which the Corporation was a net seller.

The scheme consisted of four operational elements to carry out its function: the minimum reserve price, the management of stocks, the wool tax (a levy on wool growers' sales receipts) needed to partially finance the scheme, and borrowings required when tax receipts and cash reserves were insufficient to cover the costs of stock purchases. Purchases of stocks under the scheme were funded basically by the wool tax, with no direct government funding from general revenue sources. The accumulated funds (known as the Market Support Fund) were used to purchase wool and support the market when prices were low. When the Market Support Fund was exhausted, any purchases were financed by borrowing from commercial sources against the collateral of the wool stockpile. On the other hand, any surpluses in the Market Support Fund were invested and the resultant interest returned to the Fund. When funds considered to be in excess of requirements were accumulated, these were refunded to wool growers.

Objectives of the Scheme

Any marketing scheme must have a set of objectives that its operators seek to achieve. In the case of the Reserve Price Scheme, however, the objectives were documented only in very general terms and had been subject to various interpretations (ABARE, 1990). Given the lack of clearly specified objectives, there were other more specific objectives suggested of the Reserve Price Scheme. For example, Ward (1985) suggested possible objectives of the Reserve Price Scheme as being: price and income stabilisation for producers; enhanced demand for wool via price stabilisation in user currencies; use of some market power to maximise grower returns; protection against extremely low prices; and equity between growers within and between seasons. However, it is important to recognise that to achieve these different objectives, several Policy instruments are required.

Despite the possible objectives that have been proposed for the Reserve Price Scheme, it is clear that, from a national perspective, the fundamental goal of the scheme should have been to raise the economic efficiency of the economy and, hence, the welfare of society as a whole (ABARE, 1990).

According to Newbery and Stiglitz (1981), if economic efficiency in a competitive system is to be maximised, there must be a complete set of risk markets: futures, capital and insurance markets. The fact that in reality there are uninsurable risks suggests that actual risk markets are less than complete. There may be scope for government intervention to compensate for the incompleteness in risk markets and, hence, to improve economic efficiency. Price stabilization schemes can be viewed as potentially providing such a form of intervention (Newbery and Stiglitz, 1981).

Therefore, to the extent that the Reserve Price Scheme had the effect of producing more stable wool prices, such intervention might have been justified on economic efficiency grounds if it had the effect of efficiently addressing market failures in either risk or capital markets prior to deregulation of the financial markets in Australia in 1983. In other words, the impact of the scheme on economic efficiency could be measured by the extent to which, if at all, the benefits of the scheme in stabilising wool prices (and, hence, correcting for incomplete risk markets) exceeded the costs associated with the scheme's operation (ABARE, 1990).

Costs and Benefits of the Scheme

Foster (1988) showed that the direct costs of operating the Reserve Price Scheme were considerable. The bulk of the direct operating costs were interest costs, imputed and direct, on the producer funds held to finance wool purchases (see Table 4.1). These costs were offset to some extent by gross profits made on the sale of wool from stocks. Nevertheless, over the period 1974-75 to 1989-90, the accumulated net operating losses associated with the Reserve Price Scheme totalled about $A1.7 billion when expressed in 1990-91 dollars. This represents about 2.6 per cent of the gross value of shorn wool produced over the life of the scheme to that time. A key feature of these operating costs is the major blowout in operating losses in 1989-90 as Australian Wool Corporation's stocks rose from 188,000 bales at the start of the season to a record 3 million bales by June, 1990. This increase in stocks added around $A1 billion (in 1990-91 dollars) to the net operating losses of the scheme (ABARE, 1990). Large additional losses were incurred during the 1990-91 season, before the scheme was abandoned, when Australian Wool Corporation stocks rose to around 4.7 million bales.

The key issue, then, is whether the benefits that flowed from the scheme for wool growers and the community exceeded these operating costs. The existence of benefits at several levels have been suggested for the scheme. First, there may have been benefits to purchasers of wool as a result of the scheme. Second, Australian wool growers may have benefited in several ways. For example, they may have faced less variable prices and perhaps less variable incomes as a result of the operation of the scheme. In addition, there may have been some revenue transfers to growers as a consequence of the scheme. These possibilities are examined below.

The overall impact of the Reserve Price Scheme on the demand for wool is not clear because it is not known to what extent prices and throughput were stabilised for users of wool. Although the Reserve Price Scheme may have resulted in more stable prices in Australian dollar terms for much of its life, fluctuations in the exchange rate may have meant that it did not stabilise prices in user currencies (Ward, 1985). Given that exchange rate changes are relatively independent of wool price changes in Australian dollar terms, a stable Australian dollar price of wool could either stabilise or destabilise prices in user currencies depending on the movements in the exchange rates. The variability of prices in user currencies increased following the floating of the Australian dollar in December, 1983 (ABARE, 1990). Moreover, if price fluctuations persisted in the market for finished woollen products then the profits of wool processors and retailers could have been destabilised by the Reserve Price Scheme because the scheme would tend to reduce the extent to which the price of raw wool moved in line with the price of finished woollen products. This result could reduce the attractiveness of wool to processors and retailers. Research by Quiggin (1983) provided some evidence that the profits of wool users were, in fact, destabilised by the Reserve Price Scheme.

At the industry level, studies have shown that the Reserve Price Scheme reduced the variability of wool prices (in Australian dollars) at auction (Campbell, Gardiner and Haszler, 1980; Longmire, Kaine-Jones, and Musgrave, 1986). Measured at the industry level the benefits to growers from the reduced risk of wool price volatility resulting from the scheme are likely to have been of the order of 1 per cent of the gross value of wool produced for the period until the mid-1980s (Hinchy and Fisher 1988). However, the above studies did not analyse the period of extreme prices in 1987-88 when Australian Wool Corporation's stocks were effectively exhausted nor did they analyze the effects in the period following when the levy on growers necessary to support the scheme before its collapse rose from 8 per cent to 25 per cent of gross industry revenue.

Although there is some empirical evidence that the Reserve Price Scheme may have had a significant stabilising effect on aggregate wool industry revenues (Motha, Sheales, and Saad, 1975), there is little evidence to suggest that this stabilising influence flowed through to individual growers. The Reserve Price Scheme could have only reduced instability in farm returns from wool production if a substantial component of the year to year variability in growers' returns was accounted for by corresponding changes in the price of wool.

An analysis of the relationship between wool growers' net cash incomes and wool price changes in the pastoral zone of Australia over the period 1952-53 to

1982-83 (Breckling, Fisher, and Wallace, 1989) indicated that the variability in wool prices over that period accounted for very little of the observed variability in growers' incomes. Furthermore, this result remained true both before and after the introduction of the Reserve Price Scheme in the early 1970's. Further evidence about the relationship between changes in aggregate wool prices and changes in farm level incomes suggested that a major component of the year to year variability in growers' returns was not accounted for by corresponding changes in the price of wool and that instability in farm level returns from wool production was not reduced significantly by moderating the variability of aggregate market prices (ABARE, 1990).

In addition to the impact (if any) the Reserve Price Scheme may have had on reducing the risks faced by wool producers, the scheme may also have changed the average price of wool, and, hence, affected the average level of grower incomes, a so-called revenue transfer effect. Most studies of the Reserve Price Scheme have focused on the revenue transfer effects of the scheme. While these studies have consistently shown some effect on grower revenues, the results have conflicted in terms of the estimated direction of the effect. Simmons and Arthur (1988) found that changes in producer revenue resulting from the scheme were positive but small and "unlikely to be of major significance in the evaluation of price stabilization in the Australian wool market". This is in contrast to the results obtained by Campbell, Gardiner, and Haszler (1980) and Hinchy and Fisher (1988), who concluded that there was a small loss in producer revenue due to revenue transfers. Care needs to be exercised when interpreting these results because some of them are in, fact model, dependent.

The potential benefits from the Reserve Price Scheme to wool growers need to be weighed against the operating losses of the scheme (Table 4.1) incurred by the Australian Wool Corporation over the life of the scheme. Moreover, the potential benefits also need to be weighed against other costs imposed on the rest of society through the financing arrangements of the scheme in relation to taxes and the government guarantees of borrowings that were available to the Australian Wool Corporation. For example, according to a recent study by McLachlan, Valdes and Ironfield (1991), the value of government guarantee of Australian Wool Corporation borrowings (which enables the Corporation to obtain loans at lower rates of interest) is estimated to be $A77.5 million in 1990-91. By offering government guarantees, most of the commercial risk to a lending institution on loans to a statutory corporation is removed. Moreover, the interest rates on loans to the corporation are lower than they would be without the government guarantees. When the risk of an activity is altered by offering guarantees, there is also the potential for behaviour to change, encouraging actions which would not have occurred otherwise and creating a "moral hazard" problem. There also are resource allocation effects, including the changed allocation of resources among competing farm level activities,

Table 4.1 Total direct costs of the Reserve Price Scheme to Australian wool growers[a]

	Total cost in 1990-91 dollars[b]		Cost per kilogram (clean) in 1990-91 dollars[b]		Total cost as a share of the gross value of shorn wool produced	
	1974-75 to	1974-75 to	1974-75 to	1974-75 to	1974-75 to	1974-75 to
Item	1988-89	1989-90	1988-89	1989-90	1988-89	1989-90
	$m	$m	c/kg	c/kg	Percent	Percent
Costs						
Physical[c]	429	462	6.4	6.3	0.4	0.9
Interest[d]	1,201	1,607	17.9	21.8	1.0	2.8
Total	1,630	2,069	24.3	28.1	1.4	3.7
Gross trading surplus[e]	913	336	13.6	4.6	0.8	1.1
Net operating losses	717	1,733	10.7	23.5	0.6	2.6

[a] Amounts relate to the operation of the Market Support Fund. [b] Refers to all shorn wool produced in Australia over the period, in 1990 dollars. [c] Handling, storage, selling, and administration costs. [d] Interest paid on commercial borrowings and imputed on Market Support Fund moneys tied up in wool stocks. [e] Difference between purchase cost and sale proceeds of Reserve Price Scheme wool. The drop in value between 1988-89 and 1989-90 is due primarily to the revaluation of stocks in the light of the lowering of the minimum reserve price.
Source: ABARE (1990).

because of changes in the relative risks associated with different farm enterprises.

It has also been recognised that the benefits to producers of price stabilization schemes depend on how producer price expectations are formed. Scandizzo, Hazell, and Anderson (1983) argued that if producers form their expectations rationally (that is, full use is made of all market information), then the social welfare gain from price stabilization schemes is relatively small. Fisher (1983) demonstrated that Australian wool growers may have formed their price expectations rationally, both prior to and after the introduction of the Reserve Price Scheme. He concluded that it was not possible to reject the assumption that the introduction of the Reserve Price Scheme "has had little impact on the way in which wool growers form their price expectations". It follows that the introduction of the Reserve Price Scheme is likely to have brought few benefits to producers from reduced errors in forming price expectations.

When the direct and indirect operating costs of the Reserve Price Scheme over the entire period of its operation are compared with the likely benefits, the weight of evidence is that the scheme was not of net benefit to Australian wool growers. Furthermore, there is little to suggest any gains in economic efficiency from the perspective of the Australian economy as a whole as a consequence of the existence of the scheme.

Collapse of the Wool Buffer Stock

There is a general consensus that the Reserve Price Scheme was reasonably successful in reducing the variability of wool prices during the 1970s and the early part of the 1980s. The events leading to the collapse of the wool buffer stock began in 1983-84 when the Australian Wool Corporation had accumulated a stockpile of 1.6 million bales. Because the Corporation was holding such a large volume of wool stocks, private buyers and wool users had little incentive to hold stocks and consequently allowed their stocks to run down. At about the same time, the Australian dollar was floated and Australian financial markets were deregulated. The environment in which the buffer stock scheme was being operated was, therefore, substantially different from that experienced by the Australian Wool Corporation in the past (Figures 4.1 and 4.2).

In the mid-1980's the Australian dollar fell sharply against major wool user currencies, thus enhancing wool's competitiveness against cotton and synthetics. Increased demand for wool from China, the Soviet Union, and Japan during the middle to latter part of the 1980's then resulted in a significant rundown of the wool stockpile held by the Australian Wool Corporation. In 1987-88, wool prices rose rapidly as strong demand from users to rebuild their stocks virtually eliminated the Corporation's stockpile (Figures 4.1 and 4.2). Normally, prices would have fallen when privately held stocks had been replenished. However, the reserve price was increased by the Australian Wool Corporation (with agreement by the industry) from 508c/kg clean wool in 1986-87 to 870c/kg clean wool in 1988-89 (an increase of more than 70 per cent) and held at that level in 1989-90.

The new high prices and a run of three good seasons set off a boom in Australian wool production in the late 1980's. The level of the reserve price and confidence in the Reserve Price Scheme was a significant long term factor in grower decisions (Wool Review Committee, 1991). Coinciding with this boom in wool production was a considerable decline in demand for wool in 1989-90.

In both the Soviet Union and China domestic economic problems caused a 50 per cent reduction in their wool purchases in 1989-90 as compared with 1988-89. This was in contrast to the period 1986-87 to 1988-89 when these two countries together accounted for over 20 per cent of Australian wool exports and were particularly important for wool demand in the 23-28 micron range. During

58

Figure 4.1 AWC stocks and the ratio of the market indicator to the floor price

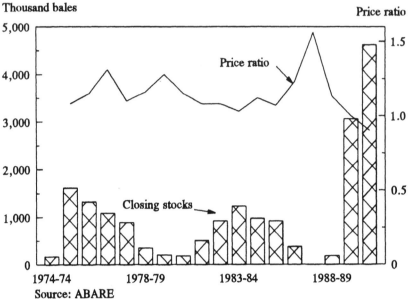

Thousand bales

Price ratio

Source: ABARE

Figure 4.2 Average price and the reserve price for wool

A$ per hundred kilograms clean wool

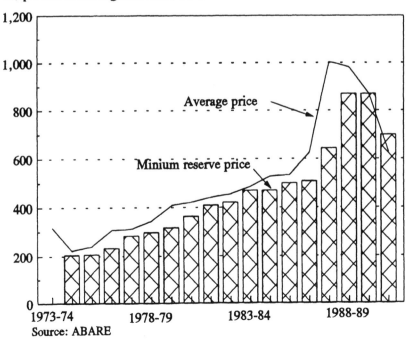

Source: ABARE

1989-90, Japanese purchases of wool also weakened because of large Japanese stocks, much of which were purchased at the high price of the previous years.

Suspension of the Reserve Price Scheme

As a consequence of much weaker demand for wool and increased supply, the Australian Wool Corporation was active throughout most of the 1989-90 season purchasing wool to maintain the reserve price. This support buying clearly sustained the auction prices well above the level that would have been determined by market forces and, in particular, above the underlying long term trend (Figure 4.3).[1] As a result, the Corporation's wool stocks rose from 188,000 bales in July 1989 to 2.45 million bales in mid-April 1990.

In May 1990, with mounting stocks held by the Australian Wool Corporation and after much debate, the reserve price was reduced, by the government using its powers under the Wool Marketing Act of 1987, from 870c/kg clean to 700c/kg clean. The reduction took effect in June 1990 for buyers and in July 1990 for wool growers. Growers who sold wool during the interim received deficiency payments from the Australian Wool Corporation. The government and wool industry leaders gave frequent assurances that there would be no further downward movements in the reserve price following the reduction to 700c/kg in May 1990. Against this background, the government established a Committee of Review into the Wool Industry in July 1990 to inquire into the effectiveness of wool marketing arrangements in Australia.

Despite the lowering of the reserve price in May 1990, the imbalance between supply and demand continued, with Corporation stocks rising to 4.6 million bales in December 1990. The new reserve price of 700c/kg proved to be too high and buyers were not confident that even this lower price could be sustained. This lack of confidence, combined with a continued high level of wool production (due to the absence of viable alternatives for many wool growers in the short term) caused wool stocks and the debt of the Corporation to rapidly reach levels, which in the view of the government, were the maximums permissible (Wool Review Committee, 1991). In February 1991, the Reserve Price Scheme was suspended for the remainder of the 1990-91 season. Wool was sold without intervention when the market was reopened.

Also, in February 1991 a supplementary payment scheme was established to ensure that those wool growers selling during the remainder of the 1990-91

---

[1]Although there is some arbitrariness about the slope of the long term trend line it is clear that even if the higher average real price of the 1960s and early 1970s is ignored the real floor price set in the late 1980s was well above the real price experience in the previous ten years.

Figure 4.3 Australian real wool prices, 1960-61 to 1989-90 in 1989-90 dollars
A$ per hundred kilograms clean wool

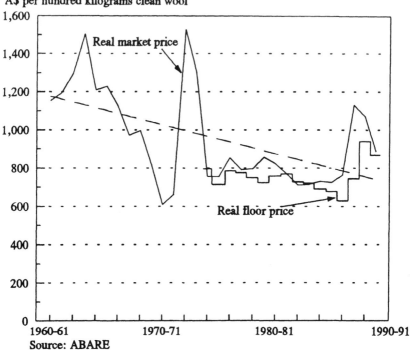

Source: ABARE

season would not be unduly disadvantaged by the suspension of the Reserve
Price Scheme. The supplementary payment scheme was jointly financed by the
wool industry through continuation of the 25 per cent wool tax and through the
government's contribution of an additional $A300 million. In April 1991, the
government decided to quarantine the Corporation's wool stocks from the market
for the balance of 1990-91. This decision was subsequently partially reversed
when around 100 000 bales of wool from the stockpile were offered for sale at
auction, of which approximately half was sold in May-June 1991, in the face of
a shortfall in the supply of certain types of wool in the market and a temporary
sharp increase in prices.

Following the completion of the inquiry into the efficiency and effectiveness
of wool marketing arrangements in March 1991 (Wool Review Committee,
1991), the government announced on 30 April 1991 a wide ranging reform
package for the wool industry, including a permanent removal of the Reserve
Price Scheme effective from the commencement of the 1991-92 season.

Return to the Free Market

Australian Wool Corporation stocks amounted to a record 4.6 million bales (800 kt) at the end of June 1991. The net present value of this asset is, of course, highly uncertain. However, most commentators agree that, on the best information currently available, the net present value of the wool will fall well short of the Australian Wool Corporation's accumulated borrowings of around $A2.6 billion. The existence of the wool stockpile and the debt imposes direct as well as indirect costs to the industry: direct costs because of any shortfall between the net present value of the stockpile and industry debt and indirect costs because of the uncertainties it creates for both buyers and sellers of wool.

In order to minimize this uncertainty and to restore confidence in the wool market, the government has given the recently established Wool Realisation Commission responsibility for disposing of the stockpile and other assets, and for managing the debt. According to the guidelines put forward by the government the debt of around $A2.6 billion is to be repaid over a maximum period of 7 years, with minimum annual debt reduction targets (Table 4.2).

Repayments are to be made through income earned from a combination of sales from the stockpile, sale of property assets originally owned by the Australian Wool Corporation, and continuation of the levy on the sale of wool (the wool tax, set at a rate of 12 percent in 1991-92). The government's guarantee of the industry's debt will continue for the repayment schedule.

The present level of wool stocks relative to current production (Figure 4.4) means that issues related to how best to dispose of the current stockpile are of crucial importance to the industry. The stockpile has both advantages and disadvantages for wool users. It provides an assured source of supply for most wool types into the future, thus reducing the levels at which private stocks need to be held. It will also effectively constrain future price rises. On the other hand, wool users fear costly instability in the market arising from any disruptive liquidation of stocks. For wool producers, the stockpile represents an asset, the sale of which will assist in repaying some or all of the debt (Wool Review Committee, 1991). However, this positive feature is countered to a significant degree by the depressing effect which sales of wool from the stockpile will have on prices and, hence, profits earned from future wool production.

There are several factors that will influence the optimal stock disposal strategy. Important amongst these are the factors that influence the net present value of the stockpile: namely, the interest or discount rate applied to the value of future sales, the physical cost of storage, and the expectations and future levels of wool demand and supply (ABARE, 1990). However, in formulating an optimal stock disposal strategy, it is also important to take into account the ownership of the wool stock.

The wool stockpile was funded by a combination of equity (the wool growers' accumulated contribution to the Market Support Fund net of funds

Figure 4.4 Supply of wool in the major producing countries
Thousand tons greasy

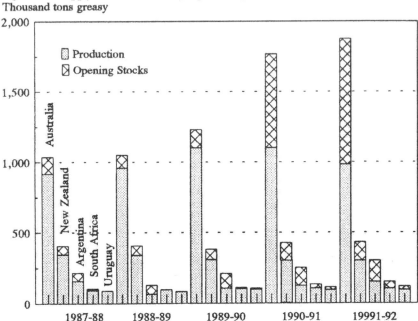

Source: ABARE

Table 4.2 Debt reduction targets

Year	Debt Reduction
	$m
1991-92	20
1992-93	300
1993-94	400
1994-95	500
1995-96	550
1996-97	550
1997-98	any remaining debt

revolved back to growers) and debt. The equity holders (in this case, wool
growers) are the owners of any assets after debt obligations are repaid. Control
is vested in the equity holders through managers who act on their behalf
(Bardsley 1991). If the Wool Realisation Commission was viewed as the

monopoly owner of the stockpile, and it was viewed as being independent of growers, then returns to the Commission from sales of stocks in a given period would be simply equal to sales revenue and the key objective of the Commission would be to maximise the net present value of these sales.

It is more reasonable, however, to view growers as the owners of the stockpile and to consider the Wool Realisation Commission as an agency acting on the behalf of growers and Australian society as a whole. In this case, the effect of stock sales on market prices, and, hence, on returns from sales of wool, must also be taken into account in determining the optimal stockpile disposal strategy. This is because sales from stocks lower the profits that growers would otherwise receive from current period production. Thus, for a given volume of sales, both the average and the marginal returns are lower for wool growers than for a private stockholding enterprise. The lower returns imply that the optimal release rate for wool growers is slower than for a private monopolist. (For further discussion of these issues, see Beare and Fisher, 1991.)

Approaches to Stock Disposal

Regardless of whether the broad objective is to maximise the net present value of the stockpile or to maximise the net present value of overall wool industry returns, there are two key alternative approaches to disposing of the stockpile. The first involves a more flexible procedure based on the strategic judgment and discretion of the stockpile manager. The aim of the second approach is to dispose of the stockpile on the basis of a set of well-defined and transparent rules related to trigger prices and quantity of stock releases (Wool Review Committee, 1991). Within these two approaches there is, of course, a range of combinations involving varying degrees of discretion among agreed rules.

The debt (and stockpile) reduction strategy currently being implemented under the direction of the Wool Realisation Commission is broadly a combination involving some discretion within a set of guidelines. The maximum period of 7 years within which the debt needs to be repaid and also the explicitly stated minimum annual debt reduction targets (see Table 4.2) attached to the repayment period can be interpreted as rules within which the Commission is required to operate. These rules broadly reflect the stock reduction strategy that is likely to be optimal from the viewpoint of maximising the net present value of industry returns, given present expectations about the supply of and demand for wool, storage costs, and interest rates. However, in managing the debt, the Commission is expected to take advantage of periods of strong demand to accelerate the repayment process. On the other hand, the Commission would be expected to use its discretion to sell only the minimum quantity of wool necessary to meet the debt reduction target at times when the market price is significantly below its medium term trend. Furthermore, the Commission is expected to provide as much transparency as possible consistent with its overall charter, including regular publication of statistics on actual sales from stocks and the level of remaining stocks.

Conclusion

There are several important lessons that can be drawn from Australia's recent experience with the collapse of its wool price stabilization scheme. When the reserve price was conservatively set (particularly during the 1970's and the early part of the 1980's), it did little more than provide a safety net for growers under extreme circumstances. It is possible that some form of price stabilization might have been justified on economic efficiency grounds to compensate for failure in insurance and capital markets. However, available evidence indicates that, in reality, the operating costs of the scheme outweighed any insurance value generated. This conclusion is further reinforced once costs other than operating costs, such as resource misallocation between competing farm industries, are taken into account. There is little evidence that the scheme was capable of stabilising wool growers' incomes at the farm level. Further, there is no unequivocal evidence that the scheme stimulated world wool demand, and the scheme may even have had adverse effects in that area. Therefore, there is little to suggest any gains from the scheme in terms of improved economic efficiency (ABARE, 1990).

Changes to the setting of the reserve price in the mid-1980's from its traditional, more conservative, level to one that was potentially interventionist largely contributed to the collapse of the Reserve Price Scheme. Once the scheme became the subject of speculative attack it was virtually inevitable that it would fail. given the importance of Australia in the world wool market and the rapid increase in wool production in Australia in the latter part of the 1980's. The consequence of that failure is that the Australian wool industry is now faced with a debt equivalent to about 90 percent of the gross value of wool production in 1991-92, valued at the expected market price of 550c/kg for that period. These estimates are based on the assumption that the Commission will dispose of about 426,000 bales (75 kt) of wool from its stocks in 1991-92 (Barrett, Tran and Roper 1991). In addition, in attempting to defend the scheme, the industry made use of reserve funds equivalent to about 65 percent of the value of current production. Further, during the period from early in 1990 until the announcement was made to suspend the scheme both wool growers and wool buyers faced a great deal of uncertainty, something the scheme was originally designed to mitigate.

Given this experience with the wool buffer stock and the experience with similar schemes for a range of other commodities, it is highly questionable whether such schemes enhance the welfare of agricultural producers in the longer term. Where there is no intervention there are a number of "stabilisation" facilities available to minimize price and income variability. For example, producers can use a variety of management techniques including the use of futures markets (Newbery, 1990). However, care needs to be exercised in the choice of stabilisation mechanism used. If, as is likely, the probability of failure of actual buffer stock schemes is close to unity and given that the financial consequences of such failure are likely to be serious, it follows that stabilisation mechanisms based on buffer stock arrangements have little place in the policy armoury of either industrialised or developing countries.

References

ABARE (Australian Bureau of Agricultural and Resource Economics)(1990), *Wool Price Stabilization,* Submission to the Wool Review Committee, AGPS, Canberra.

Bardsley, P. (1991), *Wool on the brink -- the public cost of underwriting the wool market.* Paper presented at the 35th Annual Conference of the Australian Agricultural Economics Society, University of New England, Armidale, 11-14 February.

Beare, S.C. and B. S. Fisher, and A. G. Sutcliff (1991), *Managing the Disposal of Australia's Wool Stockpile,* ABARE Technical Paper 91.2, Australian Government Publishing Service, Canberra.

Barrett, D.L., Q.T. Tran and H.E. Roper (1991), "Wool", *Agriculture and Resources Quarterly, 3(4),456-9.*

Breckling, J., B. Fisher, and N. Wallace, (1989), *An analysis of farm level income variation in the Australian wool industry: the case of the pastoral zone.* ABARE paper presented at the 1989 Australasian Meeting of the Econometrics Society, University of New England, Armidale, 13-15 July.

Campbell,K.O., and B.S. Fisher (1991), *Agriculture Marketing and Prices,* Longman Cheshire, Melborne.

Campbell, R., B. Gardiner, and H. Haszler, (1980), "On the hidden revenue effects of wool price stabilization in Australia: Initial results," *Australian Journal of Agricultural Economics* 24(1), 1-5.

Fisher, B.S. (1983), "Rational expectations in the Australian wool industry," *Australian Journal of Agricultural Economics,* 27(3), 212-20.

Fisher, B. S. and R.G. Munro, (1983), "Supply response in the Australian extensive livestock and cropping industries: a study of intentions and expectations," *Australian Journal of Agricultural Economics,* 27(1), 1-11.

Fisher, B.S. and C.A. Wall, (1990), "Supply response in the Australian sheep industry: a profit function approach," *Australian Journal of Agricultural Economics,* 34(2), 147-66.

Foster, M. (1988), *Modelling the Australian Wool Corporation's costs and revenues under the reserve price scheme for wool.* ABARE paper presented at the 32nd Annual Conference of the Australian Agricultural Economics Society, La Trobe University, Melbourne, 8-12 February.

Hinchy, M. and B. Fisher,(1988), "Benefits from price stabilization to producers and processors: the Australian buffer-stock scheme for wool", *American Journal of Agricultural Economics*, 70(3), 604-15.

Jolly, L., A. Beck, and P. Bodman (1990), *Commodity Price Stabilization in Papua New Guinea*, ABARE Discussion Paper 90.2, AGPS, Canberra.

Longmire, J.L., B.R. Brideoake, R.H. Blanks, and N.H. Hall, (1979), *A Regional Programming Model of the Grazing Industry*, BAE Occasional Paper No. 48, AGPS, Canberra.

Longmire, M.L., G. Kaine-Jones, and W. Musgrave, (1986), *Comparison of a buffer-stock scheme with a buffer-fund scheme for the wool industry*. Paper presented at the 30th Annual Conference of the Australian Agricultural Economics Society, Australian National University, Canberra, 3-5 February.

McLachlan, R., R. Valdes and D. Ironfield (1991). "Assistance to the Wool Industry", Industry Commission Working Paper No. 3, Canberra, December.

Motha, G., T.C. Sheales, and M.M. Saad, (1975), "Fluctuations in Australian rural production and prices -- some implications for support policies", *Quarterly Review of Agricultural Economics 28(1)*, 38-48.

Newbery, D.M. (1990), "Commodity price stabilization," in Scott, M. and Lal, D.J. (eds), *Public Policy and Economic Development: Essays in Honour of Ian Little*. Cambridge University Press, pp. 80-108.

Newbery, D.M. and J.E. Stiglitz, (1981), *The Theory of Commodity Price Stabilization: A Study in the Economics of Risk*. Clarendon Press, Oxford.

Quiggin, J.C. (1983), "Wool price stabilization and profit risk for wool users," *Australian Journal of Agricultural Economics*. 27(1), 41-3.

Scandizzo, P.L., B. R. Hazell, and J. R. Anderson, (1983), "Producers' price expectations and the size of the welfare gains from price stabilization", *Review of Marketing and Agricultural Economics*. 51(2), 93-107.

Simmons, P. and J. Arthur, (1988), *The importance of hidden gains and losses: a case study of Australian wool*. ABARE paper presented at the 32nd Annual Conference of the Australian Agricultural Economics Society, La Trobe University, Melbourne, 8-12 February.

Ward, L.E. (1985), "The reserve price scheme: the next ten years," *Wool Technology and Sheep Breeding*. 33(2), 45-6.

Wool Review Committee (1991), *The Australian Wool Industry, Recommendations for the Future*, Report to the Minister for Primary Industries and Energy by the Committee of Review into the Wool Industry, Canberra, March.

5

Canadian Experience with, and Reliance on, Income Security and Stabilization Measures

D. McClatchy, B. Gilmour,
J. Gellner, and B. Huff

Introduction

As we understand it, this volume is motivated by a concern that global agricultural policy changes (recent and forthcoming, unilateral and multilateral, and, in particular, a reduction in stockholding by the U.S. Government) may lead to significantly increased instability of world market prices in the future.[1] It is assumed that most governments would want to do something to correct or offset such instability, and that the principal focus of the volume is to examine the possibility of some system of international stocks management as one possible approach. Accepting the stocks management option as the main subject of discussion here, it may, nevertheless, be useful to take some time to consider whether other minimally-distortive options, that are consistent with the market-orientation objective, exist. One possibility may be the curative rather than preventative approach of Canadian-style farm income stabilization measures applied at the national level. We take it to be in this light that the subject of this

---

[1]We would note that this runs counter to the arguments of several authors that a multilateral reduction, through tariffication, in the level of insulation of domestic markets, like that of the EC, would contribute *more* stability to world prices.

chapter was chosen to be part of the volume and will try to address our assigned topic from this point of view.

In considering compensatory direct payment stabilization programs to reduce the adverse effects of market price fluctuations, as alternatives to direct interventions in the market (via stocks purchases and sales) to reduce the price fluctuations themselves, it is important to recognize that, at least historically, Canadian stabilization programs have been dual purpose. They have been support programs--transferring income to farmers--as well as stabilizing programs.[2] As with any dual purpose program, it would be inappropriate to evaluate or criticize them only on the basis of how effectively or efficiently they meet just one objective in isolation. The choice of stabilization program target variable, for example, may be more influenced by what pressure groups want to see augmented than by what they want to see stabilized.

That being said, can we be clear about precisely what governments may want to stabilize? In fact, different interest groups influencing governments may want to stabilize different things. Export interests want to maintain a continuity of supply of individual products in order to enhance their export competitiveness. Agricultural processing firms, too, may be mainly interested in stabilizing the volume of farm production for individual commodities. Short of trying to control production directly, governments can seek to generate some stability in farmers' production intentions by providing some insurance against market (price) risk and thus foster stability in relative levels of expected profitability among commodity alternatives. Much of the theoretical literature, in fact, focuses on price stabilization as a means to increased economic efficiency by offsetting the effects of two types of market failure--lack of information and incomplete risk markets (Spriggs and Van Kooten). Farm input suppliers may want to stabilize the level of aggregate farm cash income as a means of stabilizing farmers' purchasing behaviour. Regional governments, concerned about the importance of agriculture to the local economy, may opt for this stabilization target, too. Farmers themselves could be more interested in stabilizing their individual net (family) income from all (farm and off-farm) sources. Theoretical support for this alternative may exist based on the utility that farm families derive from stabilizing their consumption activities (Spriggs

---

[2]In fact, in recent years some national stabilization programs have assumed a third important role; that of a means towards reduced balkanization (disparity) in the support provided to a given commodity across provinces. Federal tripartite stabilization contributions may be withheld if the "net benefit" from all (federal plus provincial) programs in a given province exceeds a maximum for the commodity in question predetermined by agreement between all provinces and the federal government. Initially applied in the red meats area in 1990, these agreements are gradually being extended to cover other commodities.

and Van Kooten). Thus, while the range of conceivable and credible stabilization target options is quite wide, the three principal alternatives would appear to be:

- Expected commodity price (or net margin) at the farm level,
- Farm cash income (all commodities) at the sector level,
- Farm (or farm-family) net income (all commodities or sources) at the farm level.

The next section of this chapter provides an overview of the history of federal stabilization programs in Canada over recent decades. The evidence for the success of these programs as economic stabilizers is briefly reviewed and major problems encountered and lessons learned are discussed in the subsequent section. The fourth section focuses on prospects for Canadian stabilization programs in the future, and in the last section we attempt to summarize the major evolutionary trends and to draw some conclusions relevant to the topic of this volume.

Historical Developments in Canadian Stabilization Policy

Before reviewing Canadian stabilization measures that have operated over the last two decades, it is worth recalling, in the context of this volume, that our grains policies in the 1950's and 1960's did involve some elements of unilateral stocks management (by the Canadian Wheat Board, or CWB) and indirect supply control. In earlier years, the CWB farm delivery quotas did influence (limit) production to some degree, and the 1970 "LIFT" (Lower Inventories For Tomorrow) program was a kind of "set-aside" program. However, as Canada's world market share dropped, such unilateral supply control efforts became less and less appropriate. The introduction of the stabilization programs has allowed the CWB to worry less and less over time about stabilizing Prairie farm income and to increasingly focus its attention on marketing the Prairie grain crop. As transportation policy changes have improved the capacity and efficiency of the transportation system, the CWB delivery quotas are now used essentially to regulate the flow of grain through the system in an efficient manner and no longer cause back-ups in on-farm storage (which, in turn, can constrain farm production).

Stabilization is such an important element of Canadian agricultural policy that nearly all major programs have some stabilizing element to them. Thus, it is not easy to choose a subset of Canadian farm programs and say these are our stabilization programs. The two principal approaches used to achieve stabilization of effective farm prices or incomes have been (1) regulation of domestic market prices and (2) compensatory direct ("deficiency") payments. The first approach is used in the case of the "supply managed" commodities (milk, poultry, and eggs) and, through the now-discontinued "Two-Price Wheat" program, was, for a time and until recently, part of the overall stabilization package for that major commodity. However, I assume that the market price regulation approach does not meet the criteria of the options considered in this volume.

A third, relatively less important, approach used in Canada has been market price stabilization via direct intervention in the market by the Agricultural Products Board. This approach has been used mainly for some more minor commodities (especially horticultural) and will also be omitted from further consideration here. Similarly, price pooling mechanisms of the CWB and some other marketing boards, which provide some inter-temporal (within season) price stabilization, will not be discussed here. And finally, direct compensatory payments under ad hoc programs--like Disaster Assistance, Embargo Compensation, and Special Canadian Grains Program (SCGP)--have also been important in the past in stabilizing producers' returns. However, we hope to minimize the use of such measures in the future, and they, too, will be ignored in the remainder of this presentation.

What we will address here are what we will call the "mainline" stabilization programs: the Agricultural Stabilization Act (ASA) (including the relatively recent National Tripartite Stabilization Programs, NTSP); the Western Grains Stabilization Act (WGSA); Crop Insurance (CI); and the "Initial Payments" (IP) scheme for wheat and barley in the Prairies, (both of which constitute important ongoing elements of the overall stabilization package for at least some commodities).[3] It should be noted that all of these programs with the exception of IP have been recently consolidated under one single piece of enabling legislation, the Farm Income Protection Act (FIPA, 1991). Under this Act, the Western Grains Stabilization Program has been discontinued. FIPA provides enabling legislation allowing the federal government to enter into agreements with provinces to implement the new Gross Revenue Insurance Plan (GRIP) and Net Income Stabilization Account (NISA) programs. These new programs are addressed in the fourth section of the chapter. The discussion focuses in turn on red meats stabilization and on prairie crop stabilization.

Red Meats Stabilization

The Agricultural Prices Act (APA) of 1944 was perhaps the forerunner of present day stabilization programs. The measure was initiated after World War

---

[3]The Agricultural Products Cooperative Marketing Act (APCMA) works similarly to Initial Payments (IP) in providing an effective floor price protection for some non-CWB products with a cooperative marketing structure (e.g. wheat in Ontario, maple syrup).

II to guard against post-war price collapse. The legislation was clearly intended to be applied only in an emergency with few specifics on what, when and how price protection would be provided. Price support measures were applied on a case by case basis and against serious short term problems. The legislation was intended as a temporary measure to cover the period of adjustment from a wartime to a peacetime market place (Drummond et al.).[4]

The APA was replaced in 1958 with the passage of the Agriculture Stabilization Act (ASA). The new Act was in response to serious declines in farm income in the early 1950's arising in large part from depressed prices. The Act established "named" commodities that received automatic support, under prescribed conditions, at a level of 80 percent of a 10 year moving average of market prices. Hence, this legislation marked the shift from temporary emergency assistance to mandatory support. Other (non-named) commodities could also be chosen for support at a level that was flexible. Payments triggered under both the APA and the ASA were totally funded by the federal government.

The next major change in the ASA came with the amendments of 1975. Significant changes included the shortening of the support averaging period to 5 years, an increase in the safety net percentage for named commodities from 80 percent to 90 percent, and incorporation of cash production cost changes. These changes responded to the concerns that support was being eroded under the old approach because of high inflation, which was much less of a problem in the 1950's and 1960's, and by feedgrain price rises. Another change was the provision for federal/provincial cost sharing agreements. This change recognized the shared jurisdiction for agriculture but also reflected concern with the growing balkanization of Canadian agriculture production, particularly with respect to red meats.

The concern with balkanization arose because of the increasing involvement of provincial governments in stabilization and income support programs. Serious discussions were held between both levels of government and industry in an effort to harmonize stabilization programs. These discussions did not succeed, and virtually all provinces had their own individual hog stabilization programs, many had cattle programs and some, notably British Columbia and Quebec, had programs covering all major commodities by 1980.

During the early 1980's, interest was renewed in developing a national approach to stabilization for red meats. This renewed interest reflected factors such as continuing concern over production distortions, equity of support across

---

[4]A more complete summary of Canadian stabilization measures in the earlier post-World War II period would include reference to the Agricultural Products Board Act (APBA), the Canadian Wheat Board Act (CWBA), and the Prairie Farm Administration Act (PFAA).

regions, budgetary constraints of governments, and threats of countervail action. This led to the 1985 ASA amendments which allowed for tripartite plans and provided more precise criteria for what federal/provincial agreements would contain. Subsequently, in 1986, the first agreements on red meat were signed. Two provinces signed all three agreements and four other provinces signed one or two agreements. For nonparticipating provinces, the main stumbling block was the provision in the agreements to phase out only provincial stabilization programs or those which were referred to as the "top loading" programs. The focus on the "top loading" or output subsidies was based on the view of some that these types of programs were highly production distorting, while the so-called "bottom loading" or input subsidies were not. There were, however, strongly held positions among the nonparticipating provinces that the "bottom loading" programs also had to be addressed in the agreements.

In 1989, after lengthy and intense discussions, revised agreements on read meat were signed that accommodated the concerns of most provinces. The objectives of these agreements were:

- to establish a common set of rules for tripartite plans,
- to establish greater discipline and a more uniform level of support for red meat plans across the country, and
- to harmonize federal/provincial programs.

As such, the agreements included ceilings on government support from all programs which allowed for a phase-in period and higher ceilings levels for small producing provinces. In effect, the participants recognized the need to include most programs in the disciplines in order to obtain consensus. The differential ceilings recognized the higher costs of production in the small producing provinces and also that production in these provinces would not negatively affect other provinces. The agreements also contained guidelines and decision criteria for measuring net benefits and established a committee of experts and a secretariat to define and measure the level of benefits by commodity and province.

Prairie Crop Stabilization

As used here, the Prairie region refers to the CWB area: namely, the provinces of Manitoba, Saskatchewan, Alberta, the Peace River district of British Columbia, and a small part of the extreme west of Ontario.

Crop Insurance (CI), to compensate farmers for yield losses caused by natural hazards, began under the Prairie Farm Administration Act (PFAA) in 1937. It was extended and consolidated with the Crop Insurance Act (CIA) of 1959, and was subsequently amended in 1964, 1966, 1970, 1973 and 1990. The list of eligible crops has been expanding since the inception of the program and varies by province, but nearly all crops grown are covered. In the Prairies, premium costs are now shared between the producers and both levels of government, while the provincial and federal governments share the administrative costs. In most cases, farmers have a choice of which crops they will insure and of coverage at 60 percent or 70 percent of historic average yields. In Alberta and Saskatchewan, they can also decide between using their own individual average yield or an area average yield. The price at which the insured yield is valued is fixed annually before the seeding period. Participating farmers are thus protected against severe gross income losses at the individual commodity level due to yield shortfalls. CI participation rates in the Prairies, on the rough order of 75 percent, tend to be considerably higher than in other parts of Canada.

It was against this backdrop of an already existing CI program that the Federal Task Force on Agriculture recommended that a prairie stabilization program be established, and that this program be funded by participants and the federal government, in 1969. This action was taken to replace an existing set of ad hoc policies designed, in part, to stabilize price and cash income. These earlier policies included the Temporary Wheat Reserves Act acreage payments and a guaranteed minimum price for domestically consumed wheat. The proposed program was also viewed as being complementary to existing crop insurance programs: the new program would stabilize price, crop insurance would continue to stabilize yield, and theoretically the two programs combined would stabilize income (Spriggs 1985).

Several years of public debate (and one aborted parliamentary Bill) culminated in Bill C-41, which had the objective of stabilizing aggregate cash flow (net of cash costs). Stabilization was found to be generally acceptable, and the original Western Grains Stabilization Act (WGSA) came into being in April 1976. It was seen as an acknowledgement of the federal government's responsibility to provide some form of income protection to the grain-dependent Prairie economy.

The WGSP was a voluntary program, under which both grain producers and the federal government made cash contributions to an insurance fund. Its net cash flow trigger was based on calculations involving prairie-region-wide levels of average production, prices and per-unit costs for a weighted basket of crops. Payouts occurred when the net cash income for wheat, barley, oats, rye, flaxseed, canola, and mustardseed fell below the (moving) average for the previous 5 years. Thus, shifts in grain prices and cash costs as well as region-wide production declines could generate payments. By the same token, region-wide production increases could prevent payments, even in the face of price declines.

This latter feature proved to be a cause of considerable dissatisfaction with the program which led, in turn, to its being amended in 1984. The amendment

(a) changed the calculations from a calendar year to a crop year basis to more accurately reflect producers' planning cycle, and (b) added a new, alternative trigger mechanism that removed the effect of volume changes on the payout formula. Thus, from 1984, there were two triggers to the program: an aggregate net cash flow trigger and a per unit net cash flow trigger. The trigger producing the largest payout was henceforth used. Subsequent amendments were made in 1988 which (a) extended WGSA coverage to crops other than the original seven crops and (b) made levy rates adjustment more responsive to rising deficits. The WGSA was repealed in 1991 and this program has now been terminated.

Initial Payments (IP) for wheat and barley, based on per-unit levels established each year by the federal government, are primarily advance payments designed to reduce farmers' cash flow problems. Secondarily, however, they provide effective floor price guarantees for these crops. There is an important difference between the "floor" price provided by the IP mechanism and the formula-based "floor" provided by the "stabilization" programs. Initial payments are based on expected market prices and only provide protection against unexpected price declines. They are thus, in a sense, more market-oriented than the stabilization programs, which provide a safety-net price based on a moving average of historic prices, but which can nevertheless be out of line with current market conditions at the time the crop is grown. Given this difference, however, significant grains price stop-loss insurance has, in effect, been provided through the IP mechanism in some recent years.

The initial payment, less handling and transportation charges, is received by the farmer when he delivers his grain to the local elevators. His final payment, received much later, will be the difference between the initial payment and the final season-average CWB "pool" price for the crop in question (less CWB costs). If the pool price falls below the initial payment, there is no final payment and the federal government reimburses the CWB for the deficit in the pool account.

The initial payment guarantee can affect the level of any triggered payment from a corresponding stabilization program (WGSA in the past). The higher the level at which initial prices are set, the lower the payment (if any) under the corresponding stabilization program, in the event of an unexpected market price decline. An important difference, of course, is that under IP the full cost of stabilization is met by the federal government. Furthermore, it is universally available and crop specific.

For many years, CWB pool deficits tended to be rather small and infrequent. However, in the second half of the 1980's, they have tended to become more significant, reflecting the volatility of world market grain prices. The combined deficit exceeded $200 millions in 1985/86 for wheat, oats and barley. The deficit for barley exceeded $100 millions in 1986/87. The 1990/91 deficit is expected to be large for wheat.

Past Performance of Stabilization Programs

First, the impacts of the programs as economic stabilizers are discussed. Then some other important issues of concern are reviewed in no particular order. The latter part of this section draws on material in the 1990 Reports to Ministers of Agriculture of the Grains and Oilseeds Safety Net Committee (GOSNC) and the Federal-Provincial Safety Net Committee (FPSNC) as well as other sources.[5]

There can be little doubt that CI and WGSA have caused aggregate farm cash income in the Prairies to show more year-to-year stability than it otherwise would have. In some years in the 1980's, payments under one or other or both of these programs provided a significant share of total farm cash receipts (Economic Council of Canada (ECC); Spriggs, Van Kooten et al.). In the case of CI, it seems likely that this conclusion would extend to cash income at the commodity level, both in aggregate and for individual farmers. Similarly, Spriggs et al. would, in the case of WGSA, extend this conclusion to aggregate Prairie income for each of the prairie grains, because their yield changes and their price changes, respectively, tend to be positively correlated. However, it seems equally clear that at the individual farm level, WGSA, unlike CI, may not have stabilized cash incomes in many cases, either for individual crops or for farm income from all sources. The emphasis of the WGSA on stabilizing aggregate prairie-wide sectoral income rather than individual farm income has been a major criticism of the program ever since its inception, reflected in repeated calls for more disaggregated (subregional) price and yield trigger calculations (Van Kooten et al.; Fulton). Furthermore, there is the possibility that, for a farmer producing a mix of commodities, WGSA payments sparked by low grain prices (which were simultaneously stimulating livestock enterprise profits) may destabilize overall net farm income (Van Kooten et al.). The impact of CI on the stability of overall (farm level) net income on some mixed crop/livestock farms may also be questionable.

Although empirical evidence is scarce, it seems that the combination of CI and WGSA did contribute to the stability of aggregate farm net income. In the case of ASA and, later, NTSP for red meats there is even less empirical evidence. It is probably a safe bet that these two programs both increased stability of red meats enterprise net income at both aggregate (sector) and participating farm levels. However, since these commodities do not dominate farm income in any province the way grains do in the Prairies, and given the possibility of price offsets for grains, the impact of these programs on total net

---

[5]It is noteworthy that the criteria and principles for judging "safety net" programs proposed by the two committees did not include reference to their impacts as economic stabilizers.

farm income at the sector level and even (for all but the most specialized producers) at the individual participating farm level remains very much an open question.[6]

Economic stabilization impacts aside, other problems and issues have arisen with this set of stabilization programs. These programs, and most others, have been subjected to considerable scrutiny as part of the Agricultural Policy Review (APR) process which has taken place over the past 2 years. As part of this, the GOSNC and the FPSNC reports both identified some general and some program-specific problems.

One general problem of inter-provincial and inter-commodity equity is recognized to exist when different programs apply to different commodity groups and when some programs apply in only some regions. There has been a perception, for example, in eastern Canada generally and among beef producers and horticultural producers in the west, that Prairie grains farmers were getting an unfairly large slice of the assistance pie. While this and other problems may relate more to the income transfer rather than the economic stabilization nature of the programs, they are already influencing the design of the next generation of Canadian stabilization programs.

Another, related, problem is that of resource allocation distortion when programs apply to some commodities but not others. Thus the CI is thought to bias production of covered crops upward and production of livestock and "new" crops downward (ECC), and to influence production decisions on marginal land (FPSNC). The ECC had similar criticisms of WGSA, while both the GOSNC and the FPSNC were more concerned with the WSGA impact on the choice between farm-feeding and off-farm marketing of grain. This concern about possible economic distortion impacts is heightened at the present time because of the on-going GATT negotiations. The commodity-specific nature of the red-meats stabilization programs thus make them targets for similar criticism.

A further general concern about ineffective targeting and wastage arises from perceptions that some program payments were going to farmers who didn't really need help (e.g., because higher than normal yields have offset lower than normal prices, or vice versa, or because of similar offsets between commodities). Also others may not have been getting the help that they needed (e.g., under WGSA if they specialized in canola production but payments triggers were dominated by cereal prices and yields, or if local yields were low but Prairie average yields were high). Furthermore, some individual farmers were receiving several payments under a variety of different federal and provincial programs with little or no monitoring or control of the additivity of

---

[6]Still somewhat disturbing, however, is the 1984 finding of Gauthier that in only 1 of 8 years studied did overall direct government payments move in a direction opposite to realized net farm income net of government payments.

their impact. Thus the FPSNC proposed that future programs have more of a "whole farm" focus to overcome this kind of problem. The GOSNC saw lack of sub-regional targeting as a major problem with WGSA, and noted the lack of integration between the ASA and CI in the case of crops in eastern Canada.

A more controversial conclusion of the GOSNC was that existing formulae were too market oriented, allowing safety net levels to fall too quickly in response to sustained world market price declines (artificially induced by other countries' subsidy practices), which resulted in "inadequate short-term assistance to grains and oilseeds producers." As a result, ad hoc programs were developed. The FPSNC also identified problems relating to the flexibility and financial soundness of the WGSA and, also, identified a remaining lack of producer participation in the funding of some (non-NTSP) programs under ASA.

The Future

As a result of the APR and, in particular, the deliberations of the GOSNC and the FPSNC, new umbrella safety net legislation was developed, and the FIPA passed into law in 1991. This Act has brought all safety net programs for all commodities under one piece of legislation. It authorizes agreements between the Federal Government and the provinces to provide income protection for agricultural producers through CI, net revenue insurance (e.g., NTSP), gross revenue insurance (e.g., GRIP), net income stabilization (e.g., NISA) and special measures that subject to the principles defined in the act. The ASA, CIA, and WGSA have all been repealed.

The WGSP has been dropped from the stabilization arsenal. For the near future, at least, CI, IP, and NTSP are retained while, the new GRIP program replaces the WGSP for grains and oilseeds, though with a Canada-wide application. If and as the existing federal-provincial NTSP agreements are renewed, they could possibly be merged with the new GRIP program. A totally new type of program, NISA, is also being offered which, in principle, can apply to all farm commodities. However, initially (probably appropriately) for a "trial" period it will apply only to grains, oilseeds and edible horticultural crops in those provinces which have signed on.

GRIP will be financed by both levels of government and by farmers. It will provide a combination of yield and price insurance. Each spring a target revenue will be established for each participating farmer and for each commodity. The target will be a product of probable yield, seeded acreage and a safety net price. At harvest, target revenue can be compared with actual revenue, which is based on actual yield and market price in the farmers' region, to determine whether a payment is justified. Provinces are free to establish their own safety net price (below the long-term market average) and other program details, subject to national guidelines. The GRIP provides a more transparent and certain level of stop-loss protection for the crops it covers than previous stabilization programs did. Since provinces began to announce details for the 1991-92 program, the new program has come in for some criticism from the academic community and others (see, for example, Gray et al.).

NISA focuses on a farmer's income rather than on price or yield. Farmers will deposit a percentage of their qualifying sales (grains, oilseeds and edible horticultural crops initially) into their own individual trust account to which governments will also contribute. Withdrawals from that account will be permitted when either net margin or net farm income fall below predetermined levels, but only up to the amount of the account balance; accounts cannot go into deficit. Hence, NISA cannot provide immediately effective income insurance, since it takes some time for the accounts to build up. The account balance remains the property of the farmer and can be withdrawn in full after retirement from farming.

The picture for the longer term future is much less certain and will depend, in part, on the outcome of the current round of GATT negotiations. NISA is currently an important plank in the long-term plan with its features of meeting individual needs, avoiding offsetting payments, being resource and product neutral, and having a whole-farm focus. There seems to be considerable support to retain CI as an option for those farmers who do not wish to manage their own market risk insurance.

Summary of Trends and Conclusions

Over time, experience with program implementation, changing farm sector demographics, improved analysis, and changing political pressures have influenced the evolution of Canadian stabilization programs. This evolution appears to be taking the following directions, while still having some way to go:

- From full federal funding to federal/producer cost sharing and finally to federal/provincial/producer ("tripartite") cost sharing (the CI program has, of course, had these features all along).
- From many different provincial programs and national programs applying regionally, each on a commodity by commodity basis, to a few major cost-shared federal/provincial programs applying country-wide (though with regional flexibility) and to baskets (ultimately the whole basket?) of commodities (i.e., from potentially many payments per producer to one payment per producer).
- From aggregate area program accounts to individual farm accounts.
- From a commodity (price or income) focus to a whole-farm income focus.

- From programs with potential deficits and, therefore problems, of financial soundness to programs with no deficits (and, as an intermediate step, more rapid and automatic adjustment of premium or levy rates in response to any emerging account deficit situation).

Furthermore, the continuing support for CI, the dissolving support for old-style ASA programs and for WGSA, and the design of the new NISA program all point to the most important stabilization target variable becoming net income at the farm level (both by commodity and in total). Price and sectoral net income appear to have diminished in importance as stabilization targets. The development of the new programs, GRIP and NISA (particularly the latter), reflect a major advance in Canadian agricultural policy thinking. They represent considerable progress in consolidating federal and provincial government support into a minimum number of cost-shared programs, that are national in scope, and applicable and equitable across a wide range of commodities. Furthermore, they are designed, by putting more emphasis on individual whole farm accounts and records, to gradually minimize unnecessary payments when offsets occur (e.g., high price offsetting low yield, or vice versa, or cross-commodity revenue offsets) while giving individuals better protection from cumulative adversity (e.g. price and yield both depressed, or several farm enterprises simultaneously affected adversely).

Further developments of GRIP and NISA to include all commodities will make them more market oriented and resource/commodity neutral. There is the legislative flexibility in the new FIPA to do this and to cope with whatever international obligations might arise as part of a Uruguay Round agreement. There can be little doubt that NISA, once it applies to farm income from all commodities, will meet the criteria of "green" (minimally distorting) programs being developed in Geneva.

Canada's experience with previous efforts to stabilize international grains prices through price band agreements, international stocks schemes, etc., has not been very positive. Past attempts to develop viable International Wheat Agreements in the post-World War II period, and an International Grains Agreement under the GATT, are seen in Canada as having been largely unsuccessful. International commodity agreements in general are viewed with skepticism. Furthermore, we have the impression that buffer stocks programs tend to end up costing more than analysts forecast, possibly because a greater degree of omniscience on the part of the stabilizing authority is required than with underwriting or buffer fund schemes.

Finally, it would appear that political support in Canada is for programs that provide some stabilizing safety net for net incomes at the individual farm level. Achieving market price stability for some commodities will not, for various reasons (yield variability, cross-commodity offsets) suffice to ensure individual farm income security and stability. Thus, we conclude that, even if an international stocks management scheme for grains were successfully introduced and sustained, Canada would still need its own income safety net programs. On the other hand, we feel that the new generation of Canadian programs will provide our producers with efficient and effective protection against future world price volatility, were it to continue at or even exceed the level observed in the 1980's. Other countries might find that similar options would suit their needs while helping to meet their international obligations for policy reform.

References

Drummond, W.M., W.J. Anderson, and T.C. Kerr, 1966. *A Review of Agricultural Policy in Canada*, Agricultural Economics Research Council of Canada, Ottawa, June 1966.

Economic Council of Canada, 1988, *Handling the Risks: A Report on the Prairie Grain Economy*, Supply and Services Canada, Ottawa, 87 pp.

Federal-provincial Safety Net Committee, 1990, *Report to Ministers of Agriculture*, Agriculture Canada, Ottawa, June, 43 pp.

Fulton, M., 1987, "Canadian Agricultural Policy," *Canadian Journal of Agriculture Economics*. 34 (109-126), May.

Gauthier, L., 1984, *Preliminary Analysis of the Contribution of Direct Government Payments to Realized Net Farm Income*, Working Paper #5, Agric. Stats. Div., Statistics Canada, Ottawa, April.

Gellner, J., 1991, *The Evolution of the Net Benefit Concept* paper presented to the Annual Meetings of the Canadian Agricultural Economics and Farm Management Society, Fredericton, New Brunswick, July, 7 pp.

Grains and Oilseeds Safety Net Committee, 1990, *Report to Ministers of Agriculture*, Agriculture Canada, Ottawa, April 33 pp.

Gray, R., W. Weinsensel, K. Rossaasen, H. Furtan and D. Kraft, 1991, *Proposed Amendments to GRIP*, Department of Agricultural Economics, University of Saskatchewan, Saskatoon, January, 11 pp.

Spriggs, J., 1985, "Economic Analysis of the Western Grains Stabilization Program", *Canadian Journal of Agriculture Economics*. 33 (July): 209-230.

Spriggs, J., B.W. Gould and R.M. Koroluk, 1988, "Separate Crop Accounts for the Western Grain Stabilization Program", *Canadian Journal of Agriculture Economics*, 36: 443-457.

Spriggs, J. and G.C. Van Kooten, 1988, "Rationale for Government Intervention in Canadian Agriculture: A Review of Stabilization Programs", *Canadian Journal of Agriculture Economics* 36 (March) 1-21.

Van Kooten, G.C., J. Spriggs and A. Schmitz, 1989, *The Impact of Canadian Commodity Stabilization Programs on Risk Reduction and the Supply of Agricultural Commodities*, Working Paper #2/89, Agriculture Canada, Ottawa, February, 13 pp.

6.

Stabilizing Imported Food Prices for Small Developing Countries: Any Role for Commodity Futures?

*Takamas Akiyama and
Pravin K. Trivedi*

Introduction

Ability to import food at stable prices has been one of the main concerns for food-deficit developing countries. In the past, the United States has played a role of grain provider of last resort and contributed greatly to the stability of prices and supply through its large stock, which, in turn, was accumulated due to the agricultural policies of the United States. The recent trend in the world grains market has been for freer market and this trend would accelerate in the future if the current GATT negotiations on agricultural products succeed. There is some concern that world grain prices would be more volatile under a freer market condition.

Stable food prices have been an important goal in many developing countries because of the politically and socially sensitive nature of food prices. Interest in taking measures to stabilize imported food prices may increase in food-importing developing countries in view of freer world grains market. In many instances, government-controlled buffer stock schemes, and variable tariffs/subsidies have been considered. These schemes often failed mainly because their operating rules have been either inflexible or based on some ad-hoc analysis of the market.

This chapter examines the possibility of using commodity futures for the purpose of price smoothing of food imports for small developing countries. The potential for using commodity futures rather than buffer stocks for food imports and food security was previously considered by Peck (1982). Kletzer, Newbery and Wright (1990) investigated the use of futures for smoothing the consumption of primary commodity exporters. This chapter attempts a tighter integration between the recent theoretical work on the competitive storage model and the role of futures in providing greater stability in imported food prices. The empirical relevance of the storage model is a topic of continuing research, so we do not advocate its uncritical adoption. Rather we recommend the use of a framework loosely related to it. Further, the use of commodity futures for import food price stabilization depends upon certain hypotheses regarding the relation between cash (spot) and futures price. These issues are relevant to a discussion of the difficulties surrounding the routine use of commodity futures. We also provide an illustration of our approach using a simple model essentially based on the competitive storage theory, henceforth abbreviated as CST.

The reason for the focus on small food-importing developing countries is two fold. First, for such countries it can be reasonably assumed that their import demand and futures transactions would not affect world grain spot and futures prices in any significant way. (If a large position is taken by the hedging country in the futures market on either side, long or short, the country may have to pay a large risk premium.) Second, for these countries the negative impact of high world grain prices on the country's welfare is likely to be significant. Hence, stability in imported grain prices will be potentially more beneficial to developing countries.

The rest of the chapter is organized as follows: First a welfare analysis of stable imported food prices is presented. Next recent findings on spot and futures price behavior relevant to import food price stabilization are reviewed followed by empirical confirmation of these findings in the context of world wheat market. An import food grain price stabilization program using futures follows with advantages of a futures program over a buffer-stock program. The chapter concludes by suggesting topics for further research.

Welfare Implications of Import Food Price Stabilization

Because world food prices have fluctuated widely in the past, many governments have thought it necessary to take measures to stabilize imported food prices. A brief welfare analysis of stabilizing imported food prices is given here. Assume that the country (decision-maker) is risk averse and also that a significant share of its income is spent on food. Then the relationship between welfare and expenditure on food can be depicted as in Figure 6.1. The vertical axis measures negative utility or disutility. The horizontal axis is the expenditure on imported food. Note that the disutility curve is convex.

The curve implies that disutility increases significantly when the expenditure on imported food increases. Assume that expenditure on imported foods is at

Figure 6.1 The effects of expenditure stabilization on disutility

Disutility

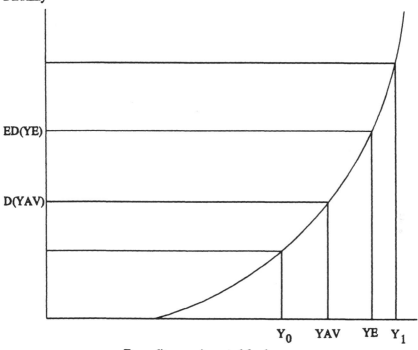

Expenditure on imported food

Y_0 and Y_1, each with probability of 0.5. In this case, the expected disutility is ED(YE), and the corresponding expenditure is YE. If the expenditure can be stabilized at YAV, the average of Y_0 and Y_1, the disutility is D(YAV), which is smaller than ED(YE). This shows that the lower the variance of Y, the lower the disutility. The difference between YE and YAV is the risk benefit and is given by:

$$RB = \frac{1}{2} R \Delta [\sigma_y^2]$$ (1)

Δ = *change (result of policy) operator*

RB = risk benefit as percentage of total expenditure on imported food.
R = coefficient of relative risk-aversion
σ_y = coefficient of variation of expenditure on imported food.
(see Chapters 6, Newbery and Stiglitz 1981).

If low price elasticity of the demand for food is assumed and if the country needs to import a certain quantity of food every year, then the higher the welfare, the lower the variance of the price of imported food. Specifically, policies that shield a country's imports from sharp price rises are said to increase national welfare.

A benefit of using futures for imported food price stabilization is clear from a simple hypothetical situation above. Under the assumptions made and with futures price assumed to be an unbiased predictor of the future spot price, the futures price should be PAV, price corresponding to YAV. Then the country could obtain imported food at PAV all the time instead of Y_0 and Y_i. This reduces the country's disutility by the amount RB in equation (1).

Spot and Futures Price Movements of Commodities

Before futures can be considered seriously as a method of providing imported food price stabilization, two issues need to be addressed. First, what is a good approximation oF the mechanism that generates food prices and what are the key insights into price behavior that such a model provides? This concerns the fundamentals of supply and demand. Second, how accurately does the futures markets reflect the influence of the fundamentals? In this section we address these issues.

There are several important studies undertaken in recent years on commodity price movements and commodity markets that are relevant to the question of imported food price stabilization. Williams and Wright (1991) provides an excellent modern account of the traditional competitive storage model and its broad empirical implications. An empirical investigations of CST is by Deaton and Laroque (1990), whereas Gersovitz and Paxson (1990) and Trivedi (1991) have studied the characteristics of commodity price movements from other perspectives. Stein (1986) presents a detailed discussion of the derivation of futures prices.

Though the empirical relevance of the CST is limited by the restrictiveness of underlying assumptions, it provides a number of useful insights into the workings of commodity markets. Three important implications of the CST for using futures to provide imported food price stabilization concern the behavior of rationally expected prices under the competitive storage model and their relationship with futures prices, the incidence of price spikes, and the expected behavior of the future variance of prices given current information.

Prices Under a Competitive Storage Model

Under the competitive storage model proposed by Williams and Wright, it is assumed that demand is stable and deterministic and also assumed that stochastic disturbances are introduced through harvests which fluctuate around a constant mean.

The basic model structure can be expressed as:

$$Z_t = \overline{Z} + v_t \qquad \text{(harvest equation)} \qquad (2)$$

$$X_t = Z_t + S_{t-1} \qquad \text{(availability equation)} \qquad (3)$$

$$S_t = g(X_t) \geq 0 \qquad \text{(competitive storagerule)} \qquad (4)$$

$$C_t = X_t - S_t \qquad \text{(consumption identity)} \qquad (5)$$

$$P_t = C_t^{-1}(P_t) \qquad \text{(price equation)} \qquad (6)$$

where

Z = actual harvest
$\overline{Z}$ = expected harvest
v_t = supply disturbance
S = stocks at end of period
C = demand for consumption
P = price
X = availability.

The system is recursive. Given Z_t, v_t, and S_{t-1}, X_t is determined. S_t is a function of X_t according to the CST,[1] and satisfies the stated nonnegativity constraint. Demand for consumption is determined by the difference between X_t and S_t. Price can be determined by finding the price that makes $C_t = X_t - S_t$.

The core of CST is the storage rule 1 for stocks which solves the following complementarity conditions:

$$p[X_t - g(X_t)] + \Gamma \geq \beta\, E p\, [Z_{t+1} + g(X_t) - g(Z_{t+1} + f(X_t))]$$

$$f(X_t) \geq 0 \qquad (7)$$

where Γ = storage cost per unit stored and β = discount factor.

There are several methods to solve this equation. For example, one method developed by Gustafson (1958) uses a linear approximation; three numerical methods are compared in Williams and Wright (appendix to chapter 3). Important characteristics of the solutions are that (a) the stock is a nonlinear function of availability, and (b) below a certain level of availability, $S_t=0$. A typical relationship between X_t and S_t is shown in Figure 6.2.

---

[1]The competitive storage rule is a solution to a complex optimizing problem. For this chapter, however, the details of a computational algorithm are not required.

Figure 6.2 Relationship between availability of X and stocks, $f(x_t)$

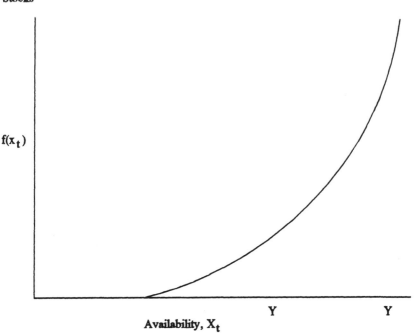

The distribution of price, denoted $f(P_t \mid v_t, S_{t-1})$, can also be generated for a given distribution of v_t. To illustrate this, we now examine the rationally expected price for period t at the end of period t-1. Assuming v_t to be normally distributed with mean of zero, the expected distribution of Z_t is depicted at the bottom of Figure 6.3 on the left. This is also the distribution of X_t when $S_{t-1} = 0$. If there is carry-in stock, the distribution of X_t is Z_t shifted to the right by S_{t-1} and is also depicted at the bottom of Figure 6.3 on the right. Demand curve is shown with and without storage. A demand curve with constant elasticity is assumed here.[2] Distribution of prices with and without S_{t-1} is shown along the vertical axis of Figure 6.3. Given S_{t-1} and the distribution of Z_t, rationally expected price is the probability weighted average of the distribution, which is the mathematical (conditional) expectation, $E(P_t \mid \Omega_{t-1})$. Note that the distribution of price is not normal (gaussian) distribution but skewed to the right.

---

[2]Effects of the functional forms for the demand curve on the model is discussed in Wright and Williams (1982).

Under the "efficient market hypothesis," i.e. assuming that learning is complete and that there is zero risk premium,[3] the futures price for period t, $F_{t+1,t}$, is an unbiased predictor of the actual (spot) price at t for P_{t+1}; i.e., $F_{t+1,t}$ = $E[P_{t+1} \mid \Omega_t]$ (Samuelson 1965, Working 1949).[4]

$$\partial E(P_{t+1} \mid P_t)/\partial P_t \geq 0 \; for \; P_t \leq P*_t$$

$$\partial E(P_{t+1} \mid P_t)/\partial P_t = 0 \; for \; P_t > P*_t \tag{8}$$

The CST also implies that where P^* denotes the price at which the storage becomes zero. In terms of S_t

$$\partial E(P_{t+1} \mid S_t)/\partial S_t < 0 \; for \; S_t > 0. \tag{9}$$

Figure 6.4 shows (7) and (8) graphically. Equations (7) and (8) imply that the expected price for t+1 increases with P_t up to the level of $P*_t$, but is constant beyond that point. (Of course in practice $P*_t$ is unknown. Deaton and Laroque treat it as a free parameter to be estimated.) To see the importance of this finding, assume a situation where P_t is above $P*_t$, but only moderately high, and the expected price is $P*_{t+1}$, which can be calculated from Figure 6.3 using the (assumed) known distribution of v_t. However, when P_t is above P^*_t but also very high, the prediction of the theory is for the same price as in the previous case.

The Behavior of the Variance of Price

We assume that the first moment of $f(P)$ exists, and consider the variability around the mean level.[5] CST (see Deaton and Laroque 1990) predicts that the

---

[3]"The magnitude of the risk premium depends upon net hedging pressure and the 'adequacy of speculation'" (p. 164, Stein). Net hedging pressure occurs if there is an imbalance between long and short position of hedgers. If this pressure is not accommodated by speculators, then the futures price is a discounted value of the market-expected future spot price.

[4]Some writers refer to the case of complete learning, or the absence of learning errors, as the case of zero "Bayesian errors", e.g. Stein (1986).

[5]It is a fact that it is difficult to reject the hypothesis of unit root in the real commodity price series. This fact is consistent with the series being a random walk and, hence, with the first moment of the data not existing. It is also consistent with trend stationarity of the series. Deaton and Laroque (1990) and Trivedi (1991) argue against the random walk model for real commodity prices.

88

Figure 6.3 Production, availability and expected price

Price

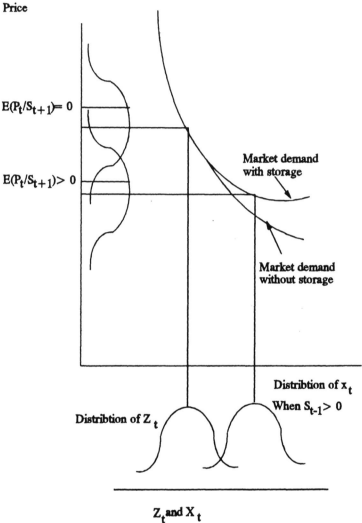

effective demand becomes much more price elastic when prices are low than when prices are high because at low prices storage absorbs availability but storage ceases to affect the effective demand when p > p*. This implies that:

$$\partial V(P_{t+1} | P_t) / \partial P_t \geq 0 \tag{10}$$

or

$$\partial V(P_{t+1} | S_t) / \partial S_t \leq 0 \tag{11}$$

The equations above imply that the variance of price (v) in the next period is greater when the current price is higher or the current stock level is lower. Empirical studies usually report considerable heteroskedasticity in the commodity prices, which is consistent with the theoretical prediction. The rational expectations hypothesis, together with the efficient market hypothesis, implies that the conditional distribution of spot and futures prices should coincide. This, in turn, suggests that the range within which the future price will fall become wider with the level of price, at least up to a certain point. Correspondingly, the probability of a high price increase inversely with the level of the carry-in stock.

Price Spikes

Many commodity real price movements are characterized by low prices for extended periods, punctuated by spikes and thus the occasional occurrence of very high prices, which are central to the discussion of food security. An important issue is the source of the spikes.[6] In the CST, the explanation lies in the asymmetric response of prices to gluts and extreme scarcity. Storage can accommodate gluts, but the non-negativity constraint on storage implies that stock-outs cannot be so accommodated. Hence, in the stock-out, situation prices will rise more sharply than they fall in periods of relative abundance.

Deaton and Laroque (1990) have shown that the resulting behavior of prices can be expressed as a switching first-order autoregression in which the autoregressive parameter becomes zero when a stock-out occurs, and prices rise above the level P*.

When stocks are positive, we have a first-order autoregression whose parameter is determined jointly by the discount rate of the holder and the "depreciation" rate for the stocks. That is:

---

[6]One possible explanation lies in competitive speculation. An oft-cited study that discusses the role of speculation in the 1972-75 commodity boom is Cooper and Lawrence (1975).

Figure 6.4 Current price, stock and expected price

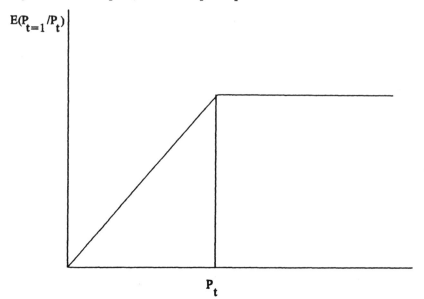

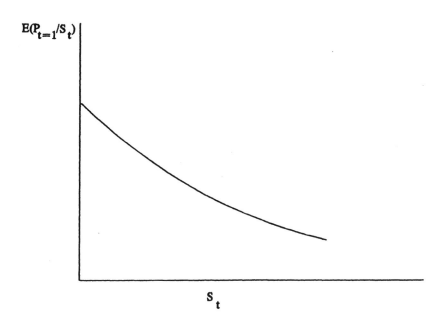

$$P_{t+1} = f(Z_{t+1}) \quad \text{if} \quad P_t \geq P_t^* \tag{12}$$

$$P_{t+1} = [\beta(1-\gamma)]^{-1} P_t + \epsilon_{t+1} \quad \text{if} \quad P_t < P_t^* \tag{13}$$

where

$$P_t^* = \beta(1-\gamma) E[f(Z_t)] \tag{14}$$

where Z = harvest

β = discount factor $1/(1+r)$

γ = "depreciation" factor reflecting per period loss in storage

ϵ = stochastic disturbance.

From (11)-(13), it is evident that prices will be serially correlated as long as $P_t < P_t^*$; that is, if the prices movements are explained by (12). When a stock-out occurs, or equivalently $P_t > P_t^*$, prices cease to depend upon the lagged prices.

The probability of price spikes is inversely related to the level of the carry-in stock. This can be explained by a hypothetical situation. Suppose there were a carry-in stock of S_{t-1} and harvest of Z_t. Assume that a price spike occurs when $X_t \equiv Z_t + S_{t-1} < X_t^s$. Then the probability of the occurrence of a price spike is given by:

$$Pr[price\ spike] = Pr[X_t < X_t^s \mid S_{t-1}]. \tag{15}$$

Substituting $X_t = \overline{Z} + v_t + S_{t-1}$ into (14), we obtain:

$$Pr[price\ spike] = Pr[\overline{Z} + v_t + S_{t-1} < X_t^s]$$

$$= Pr[v_t < X_t^s - \overline{Z} - S_{t-1}]. \tag{16}$$

It is clear that the probability of price spike declines with S_{t-1}.

Empirical Confirmation

The objective of this section is to empirically verify some of the important predictions of the CST in the context of the world grains market. Strictly speaking, however, the theoretical version of CST is not appropriate for our kind of empirical testing in a nonstationary environment. Therefore, our use of the term "CST based" should be interpreted somewhat liberally. Several other issues arise in implementing this exercise. For example: which grains should be included? What deflator should be used for nominal prices? Why do fairly substantial stocks of grains exist even in times of high prices? How should one handle government interventions in the market?

Consequently, we have restricted attention to a small model of the world wheat market based on world production, consumption, and stock data published by the United States Department Agriculture. We use wheat (No.2 soft red winter) cash price at Chicago. In view of the importance of the United States in the world wheat market, the deflator is (as in Deaton and Laroque 1990) the U.S. consumer price index.

The Relation Between Actual and Optimal Stocks

A key issue is whether actual stocks behave like the "optimal" stocks as generated by our variant of CST. To generate a series for optimal stocks, we used a model with the same structure as the theoretical model of the previous section except that the average harvest is taken as the predicted value from the regression of log (world production) on a linear trend. We found little evidence of positive supply response at the world level. The histogram of residuals from the regression suggested that the errors were approximately uniformly distributed in the range minus 7 percent to plus 7 percent. The demand equation is a log-linear function of the real wheat price, Gross Domestic Product (GDP) of the OECD countries (which proxies for the incomes of all wheat consumers) and a time trend which is a proxy for population growth and taste change, if any. Price and income elasticities were found to be -0.13 and 0.22, respectively.

Given the above model, the computation of optimal stocks can be carried out, year by year from 1965 to 1989,[7] using initial availability, and the following formulae adapted from Newbery and Stiglitz[8] (1981, chapter 30) for the case of the discrete uniform error distribution with probability mass at discrete points $Z \pm ku$, where k is a multiplicative factor:

$$S(X_t) = \alpha_t(X_t - X_{0,t}) \tag{17}$$

$$\alpha_t = \frac{1 + \beta_t - \sqrt{1 + \beta_t^2}}{\beta_t} \tag{18}$$

---

[7]The data refer to crop year, from July to June of the following year.

[8]Computations were also carried out using the assumption of normally distributed harvest shocks with a zero mean and standard deviation of 4 percent. But the results, while qualitatively similar to those reported below, were marginally inferior in terms of the goodness of fit of the stock equation reported below.

$$X_{0,t} = \frac{a_t - \frac{1}{2}\,\alpha_t\beta_t(1+3u/2)}{1 - \alpha_t\beta_t/2} \tag{19}$$

$$a_t = 1 + \epsilon_d(1 - \beta_t - \Gamma/\bar{p}) \tag{20}$$

where ϵ_d denotes price elasticity of demand, Γ denotes storage costs, and p-bar denotes average price. The term $3u/2$ was set to zero as it was quite small. The ratio Γ/p was taken to be 0.06. The real discount rate β is the U.S. prime lending rate less inflation rate measured by the Consumer Price Index (CPI). We note that the Gustafson optimal storage rule applies to a stationary environment[9] whereas we allow the discount rate and the mean harvest to vary over time, the latter in a deterministic fashion.

To see the relation between the computed optimal stocks and actual stocks, we regressed actual end-of-season stocks (SW) on an intercept, optimal stocks (SW^{opt}) and availability (AW = production + carry-in stock). The inclusion of AW is intended to take account of working stocks, not included in the competitive storage model. The estimated regression, without the intercept, which was found to be statistically insignificant,[10] is given below:

$$SW_t = 1.052\ SW_t^{opt} + 0.090\ AW_t$$
$$\quad\quad (9.79) \quad\quad\ \ (7.62) \tag{21}$$

$$R^2 = 0.922 \quad\quad SEE = 8.51 \quad\quad DW = 0.84$$

The overall fit of the equation is good and supportive of our version of CST, subject to the qualification that the computed series SW^{opt} embodies the benefit of hindsight insofar as its computation is partly based on an estimate of trend production generated using the entire sample as well as the actual data on initial availability, AW, and the GDP of the OECD countries. The residuals from the equation appeared to be slightly heteroskedastic and serially correlated, so the reported t-ratios embody the Newey-West correction. The coefficient of SW^{opt} is not statistically different from unity at 5 percent significance level. The equation says that actual stock of wheat is roughly equal to the computed optimal speculative stock plus 9 percent of the total availability, which we interpret as working stock. Thus, the estimated equation has a plausible interpretation as well as a good fit. It does overpredict during the commodity boom of 1972-73, a phenomenon that is consistent with a large speculative commodity demand at

---

[9] We thank Angus Deaton for drawing our attention to this point.

[10] When estimated with an intercept term, the latter was estimated as 1.75 with a "t-ratio" of 0.2, and, hence, was omitted from the regression.

this time (Cooper and Lawrence 1975). We shall return to this issue later in the chapter.

An important implication of this result is that, in spite of government intervention in the wheat market, the aggregate wheat stock was, on average, in line with the predictions of the CST. Another interpretation of this result is that there is a high degree of substitutability between government and private stocks of wheat: a reduction in one component is matched by an increase in the other.

Expected Price Series Generated by the Model

By solving the model, conditional on the carry-in stock, SW_{t-1}, the expected production, AW_t, predicted by the production equation, can be generated. The storage rule then provides an estimate of SW_t using (20). Subtracting expected availability, we obtain an estimate of consumption, which is inverted to solve for the price. To calculate *expected price*, we ran seven simulations with supply disturbances of -6 percent, -4 percent, -2 percent, 0 percent, +2 percent, +4 percent and +6 percent, and computed an average price under each simulation since we assume that the supply shocks are uniformly distributed.

The series of expected price (PE) is plotted against the actual price (PRWHR2) in Figure 6.5. The generated series, which under conditions of the efficient market hypothesis (which preclude learning errors and a nonzero risk premium) should mimic the futures price, is more stable than the actual spot price series. This stability of the expected price is evident when real spot prices were very high during 1973-75. The mean and standard deviation of PE are 363 and 89.8, respectively, compared with the values of 375 and 161 for PRWHR2. The simple correlation between the two is 0.83.

The CST predicts that the expected price of the following year would be inversely related to the initial carry-in stock. After some experimentation we estimated the following regression of log of real wheat price (PRW_t) on an intercept, a linear trend and the log of carry-in stock which, in our sample, is always positive:

$$\log PRW_t = 9.24 - 0.026\ t - 1.14\ \log\ SW_{t-1}.$$
$$\qquad\qquad (23.96)\quad (4.21)\qquad (5.83)\qquad\qquad\qquad (22)$$
$$R^2 = 0.59 \qquad DW = 1.21 \qquad Sample\ period = 1967-88$$

Once again, the reported absolute t-ratios are Newey-West adjusted, since the least-squares residuals show some signs of first-order serial correlation and also some heteroskedasticity (as predicted by the CST). In 1973 and 1974, and again in 1978, 1979, and 1980, the equation overpredicts. While the R^2 is only 0.6, the coefficients are very well determined and again in line with the predictions of the CST. The negative coefficient on the trend can be interpreted as reflecting the effect of choosing the deflator CPI that was growing on average 2.6 percent faster than the average world cost of wheat production.

Figure 6.5 CST model-based expected and spot prices

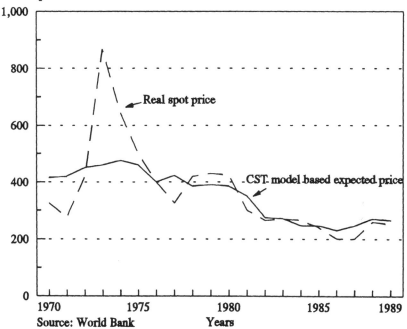

Dollars per ton

Source: World Bank   Years

The Relation Between P_t and $Var(P_{t+1})$

A test of the CST based on the conditional variance function requires a larger sample than used elsewhere in this chapter. The reason is that the prediction of the theory that $Var(P_{t+1} \mid P_t)$ is an increasing function of P_t up to a certain point, and constant beyond that, can be best tested using a nonparametric regression, or a similar technique, which allows the partial derivatives of the regression function to vary over the sample points. But implementation of such a technique requires more data points. We also require residuals from an estimated price equation consistent with the CST. For this purpose, we use the results of Deaton and Laroque, who have estimated a wheat price equation based on $(11) \sim (13)$ using annual data for the period 1900-88. Using the same data and reported parameter estimates as theirs, we were able to reconstruct the residuals from their equations reasonably precisely. Regression of squared estimated residuals on lagged price produced a coefficient of 0.0344 with a heteroskedasticity consistent t-ratio of 1.99. While the estimated positive slope is in conformity with the CST, the latter also predicts a nonconstant slope coefficient.

To examine possible nonconstancy of the slope coefficient, a nonparametric regression, which allows for local variations in the slope, was then run with squared residuals as the dependent variable and with lagged real price as the independent variable.[11] The nonparametric regression requires the investigator to make a choice of a smoothing parameter, called "group radius," or GR. The choice of GR is usually done by trial and error. For larger values of the smoothing parameter, the results approach the least squares. In the case of results reported here, the qualitative results are essentially insensitive to values of GR between 2.8 and 3.9. Values larger than 3.9 essentially reproduce the least squares regression, mentioned earlier, which supports the hypothesis that conditional variance is an increasing function of lagged price.

Figure 6.6 gives the graphic representation of the estimated nonparametric regression. It shows that the regression supports the hypothesis that variability is an increasing function of the lagged price level and also that the partial derivative of the regression increases at high prices. However, It declines at still higher prices. Given the relatively low frequency of price spikes, this latter result is subject to a wide confidence interval, as shown on the diagram.

Nonparametric Regression

Imported Grain Price Stabilization

The empirical results of the previous section suggest that a model of the world wheat market based on the competitive storage theory provides a sound basis for empirical analysis. The conditional expected price path generated by this model is also the path of the futures price, under the assumptions of rational expectations and efficient markets. A futures program that exploits the insights of the CST would contribute greatly to stabilization of imported grain prices. A "futures program" means an operation under which a small food-importing country buys grains futures contracts to hedge prices of grains it imports. It is assumed that the country always hedges with futures having delivery dates of the following year for all the import requirement before the supply disturbance for the coming harvest is known to the market. This program can be contrasted with a program under which the country buys grains spot (spot program). The futures program would be judged useful for stabilizing imported grain prices if the program stabilizes prices of the grains in the importing country at a lower level compared with prices under the spot program, especially during periods of high world prices (spikes).

---

[11]For estimation we used the N-kernel (NK) program developed by MacQueen and reviewed by Delgado and Stengos (1990).

Figure 6.6 Nonparametric regression of variability on lagged prices

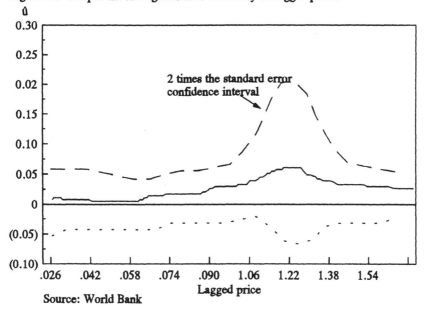

Source: World Bank

To illustrate the hypothetical performance of two programs in times of high prices, consider the following scenarios.

Suppose the carry-in stock, S_{t-1}, is adequate but the negative supply shock is large and the price increases sharply. Figure 6.7 depicts this situation. At t-1, before the information on poor harvest is known, the expected availability for t is X_t with probability distribution shown at the bottom of the figure. Assume that actual availability in t is very low at X_t^a due to very low actual harvest in t. Then the actual price in period t would be high as shown in the figure. The price of futures, maturing in t, at t-1, is $E(p_t \mid \Omega_{t-1})$, where Ω_{t-1} is all the information available at t-1, as discussed earlier.

At period t, the expected price for t+1 is $E(p_{t+1} \mid \Omega_t)$ as shown in Figure 6.6. This is higher than $E(p_t \mid \Omega_{t-1})$ because carry-in stock S_t is zero and, hence, the expected availability for t+1, $E(X_{t+1})$ is smaller than $E(X_t)$. Because supply is usually responsive to the previous year's price, $E(Z_{t+1} \mid \Omega_t)$ is higher than $E(Z_t \mid \Omega_{t-1})$. Assuming no serial correlation in harvest shocks, $E(p_{t+1} \mid \Omega_t)$ will be smaller than p_t.

However, a mechanical use of a futures program for stabilizing imported grain prices presumes that the grains futures market operates in accordance with the efficient market hypothesis. Relaxing this assumption, we may express the difference between the rationally expected and futures price as follows: equals the learning error plus the risk premium where we allow for the possibility that there is learning. Subjective rational expectations, with

98

Figure 6.7 Expected price before and during a price spike

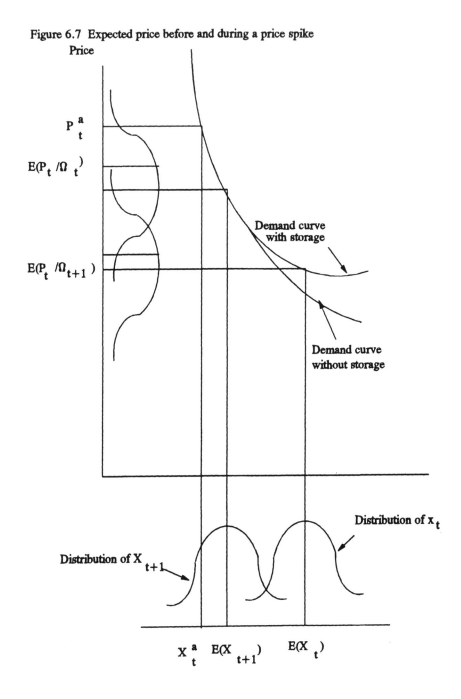

$$E[P_{t+1}|\Omega_{1,t}] - F_{t+1,t} = (\ E[P_{t+1}\ |\ \Omega_{1,t}] - E[P_{t+1}\ |\ \Omega_{2,t}])$$ (23)

$$+(E[P_{t+1}\ |\ \Omega_{2,t}] - F_{t+1,t})$$

incomplete learning, are based on the information set $\Omega_{1,t}$ that differs from $\Omega_{2,t}$ which reflects full learning. Also a nonzero risk premium exists. The relationships among the futures price, spot price, learning deviation and risk premium are shown in Figure 6.8 as the time paths of the futures price with the delivery date of T, spot price, $E[P_{t+1}|\Omega_{1,t}]$ and $E[P_{t+1}|\Omega_{2,t}]$. At T, these prices will converge, assuming zero basis. Divergences between the futures price, $E[P_{t+1}|\Omega_{1,t}]$ and $E[P_{t+1}|\Omega_{2,t}]$, can occur due to learning error and risk premium at that time. In general, the learning error and risk premium are not constant but time-variant, and can be positive or negative.

The decomposition in (23) allows us to see how the CST based model can be helpful. Suppose that $E[P_{t+1}\ |\ \Omega_{1,t}]$ refers to the prediction generated by the CST based model. The left-hand side of (23) then becomes observable. Then, additional information should be sought to determine whether the observable difference is attributable wholly to the risk premium. If the difference is large and negative, then the futures program should not be used at that time, because, in this case, the likelihood that a subsequently realized spot price will be lower than the futures price is high. On the other hand, if the difference is large and positive, the futures program would benefit by taking a long position. The difference may be large and negative because of incomplete learning by the market, in which case the difference will shrink with as the delivery date approaches.

Given historical episodes in which the futures markets behaved in a way that is apparently inconsistent with market efficiency, some investigators have suggested that speculative price bubbles may have occurred, even though the empirical evidence on these phenomena is scanty at best.[12] The existence of significant speculative movements in commodity futures is an indication of incomplete learning in the market. A CST based model should have a useful role in decision-making for a futures program. Particularly, for evaluating the divergence between the price predictions based on fundamentals and those based on the commodity futures.

The main objective of this chapter has been to achieve stability of imported food prices, for small developing countries, using a futures program with the aid

---

[12]A widely quoted study of the 1972-75 commodity boom by Cooper and Lawrence (1975) mentions speculation as one of the possible causes of high prices, but the evidence is mostly indirect and circumstantial, and not backed up by studies of volume and price variations on the futures markets.

Figure 6.8 Hypothetical movements of spot, futures, and expected prices
Price

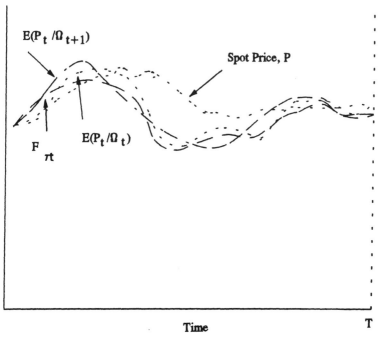

of the CST based model. Such a model is an efficient information processing and decision-making tool. Indeed, within this framework, we can also study econometrically those time-varying factors that systematically affect the futures price.

Limited research has been undertaken on using futures contracts for the purpose of stabilizations of imported food prices. Larson and Coleman (1991) examined the use of options for commodity price stabilization schemes and Peck (1982) discussed practical aspects of using futures, but neither study investigated the implications of the variability of food import prices in light of the characteristics of commodity price movements discussed above. Also, Kletzer, Newbery, and Wright (1990) have considered the use of futures rollover for longer term price stabilization, but they do not consider the choice of contract maturity.

Conclusion

This chapter has concentrated on a variant of the competitive storage model. The analysis suggests that the futures program should fare better than the spot program because the variance of prices is reduced when the futures prices are rationally expected prices, and hedging can be done before the supply disturbances are known to the market. Turnovsky (1983) has considered in some detail the distribution of futures and spot prices in an economy with consumers, risk-neutral producers, and speculators but without a nonnegativity constraint on stocks. His comparison of these distributions shows that both in the short run and the long run the futures price is more stable than the spot price, the difference in their respective variances being larger the greater is the responsiveness of supply to the futures price. These results also suggest that a futures program will stabilize expenditures relative to the spot program. However, Turnovsky's analysis is not designed to examine the choice of long versus short maturities. While there are several unresolved research issues here, the available analyses suggest that a futures program should be useful in avoiding very high prices that occur occasionally in commodity markets.

We believe that the futures program is far superior to a government-managed buffer stock scheme for small, food-importing developing countries for several reasons,[13] including the following:

- The buffer stocks will not be effective unless borders are tightly controlled. The analysis in this chapter strongly suggest that the world wheat stocks are already held in an approximately efficient way. Hence, additional buffer stock would not be effective in stabilizing wheat prices unless borders are shut to isolate the domestic market from the world market. The futures program would need tightly controlled borders if the achieved stable wheat prices are passed on to consumers directly, but the program could pass the benefit of stable prices through adjustments in taxes, especially to the poor.

- The buffer stocks would be much less flexible and speedy in adjusting to the short run market conditions. They will require purchases or sales of the physical commodity, transportation, and warehousing, any of which would cause long delays in adjusting to the buffer stocks to an appropriate level. The futures program would be free of these problems and adjustments would be made in a matter of seconds. Further, note that the cost of storage imbedded in the futures price is usually the most efficient marginal cost which usually would not be the case for costs of storage in developing countries.

---

[13]Hughes-Hallett and Ramanujam (1989) have compared the use of futures as an alternative to price stabilization via buffer stocks using a setting and an approach different from the one used in this paper.

• Many buffer stock schemes failed in the past because their operating rules were inflexible and, hence, did not adjust to changing world market conditions in the medium to long run. Because the futures price (most of the time) incorporates the key market information, it would almost automatically force the operating rules of an adequately designed futures program to be flexible and to follow the changes in the world market.

Several important issues need further investigation. Although the behavior of futures prices and their relationship with spot prices is central to a futures program, this issue was not investigated and has been left for future research. Also relevant are issues of the size and variability of risk premia (Dusak 1973; Hazuka 1984; Jagannathan 1985; Kaminsky and Kumar 1990a, 1990b; Stein 1986; and Williams 1986) as they affect the cost effectiveness of a futures program. Since *a priori* theory suggests that the behavior of risk premia may depend on the price level in a complex way, empirical investigation of these issues may also benefit from the insights of the competitive storage model. The frequency with which speculative bubble-type phenomena affect commodities futures and whether they would limit the scope for risk management is another issue deserving further examination.[14]

References

Bodie, Z., and V. Rosansky (1980) "Risk and return in commodity futures," *Financial Analysts Journal*, 36, 17-39.

Cooper, R. N., and R. Z. Lawrence (1975) "The 1972-75 commodity boom," *Brookings Papers in Economic Activity*, 671-723.

Deaton, A., and G. Laroque (1990) "On the behavior of commodity prices," pre-print, Princeton University; forthcoming *Review of Economic Studies*.

Delgado, M., and T. Stengos (1990) "N-Kernel: A review," *Journal of Applied Econometrics*, 5, 209-304. Dusak, K. (1973) "Futures trading and investor returns: an investigation into the commodity market risk premiums," *Journal of Political Economy*, 81, 1387-1406.

Fama, E. F., and K. R. French (1987) "Commodity futures prices: some evidence on forecast power, premiums, and the theory of storage," *Journal of Business*, 60.

---

[14]Systematic study of speculative bubbles based on modern analytical tools is a relatively new topic. Some recent surveys are by Shleifer and Summers (1990) and Flood and Hodrick (1990) in the *Journal of Economic Perspectives*.

Flood, R. P., and R. J. Hodrick (1990) "On testing speculative bubbles," *Journal of Economic Perspectives*, 4(2), 85-101.

Gersovitz, M., and C. Paxson (1990) *The Economies of Africa and the Prices of Their Exports*, Princeton University, Princeton Studies in International Finance, No.68.

Gustafson, R. L. (1958) *Carry-over Levels for Grains*, USDA, Technical Bulletin 1178.

Hazuka, T. (1984) "Consumption betas and backwardation in commodity markets," *Journal of Finance*, 39, 647-655.

Hughes-Hallett, A. J. and P. Ramanujam (1989) "The role of futures market as stabilizers of commodity earnings," in L. Alan Winters and David Sapsford, editors, *Primary Commodity Prices: Economic Models and Policy*, Centre for Economic Policy Research, Cambridge.

Jagannathan, R. (1985) "An investigation of commodity futures prices using the consumption-based inter-temporal pricing model," *Journal of Finance*, 40, 175-191.

Kaminsky, G., and M. S. Kumar (1990a) *Efficiency in commodity futures markets*, IMF Staff Papers, 37, 670-699.

Kaminsky, G., and M. S. Kumar (1990b) *Time varying risk premia in the futures markets*, IMF Working Paper.

Kletzer, K. M., D. M. G. Newbery, and B. D. Wright (1990) *Smoothing the consumption of primary commodity exporters: an assessment of some alternative instruments*, PRE Working Paper, World Bank.

MacQueen, D., *N-Kernel User's Manual*, Santa-Monica: N-SSS.

Newbery, D. M. G., and J. Stiglitz (1981) *The Theory of Commodity Price Stabilization*, Oxford University Press.

Newey, W. K. and K. D. West (1987) "Simple, positive definite heteroskedasticity and autocorrelation consistent covariance matrix," *Econometrica*, 85.

Peck, A. E. (1982) *Futures markets, food imports and food security*, AGREP Division Working Paper No. 43, World Bank.

Samuelson, P. A. (1965) "Proof that properly anticipated prices fluctuate randomly," *Industrial Management Review*, 6, 41-49.

Shleifer, A., and L. H. Summers (1990) "The noise trader approach to finance," *Journal of Economic Perspectives*, 4(2), 19-33.

Stein, J. L. (1986) *The Economics of Futures Markets*, Basil Blackwell.

Trivedi, P. K. (1991) *Time series analysis of commodity prices*, preprint.

Turnovsky, S. J. (1983) "The determination of spot and futures prices with storable commodities," *Econometrica*, 51, 1364-1387.

Williams, J. C. (1986) *The Economic Function of Futures Markets*, Cambridge University Press.

Williams, J. C. and B. D. Wright (1991) *Storage and Commodity Markets*. Cambridge University Press.

Wright, B. D., and J. C. Williams (1982) "The economic role of commodity storage," *Economic Journal*, 94, 596-614.

Wright, B. D., and J. C. Williams (1984) "The welfare effects of the introduction of storage," *Quarterly Journal of Economics*, 104, 275-298.

Working, H. (1949) "The investigation of economic expectations," *American Economic Review*, 39, 150-166.

7

Government Grain Storage:
Food Security and Price Stability

Robert D. Reinsel

In the previous chapters, we have been reviewing the reasons for government storage of commodities and the possibilities for alternative food security measures in unregulated markets. We will now come to closure on our discussion.

The current world economic environment makes clear that national boundaries hold less and less significance for markets. Multinational corporations move capital to labor and resources. Immigration policies are allowing labor to seek out resources and capital. Trade barriers are being reduced, currencies are being made convertible, and exchange rates are allowed to float. These events make protection of domestic agriculture and self-sufficiency goals difficult to justify.

Price enhancement for income support was the primary goal of U.S. farm programs for most years from 1933 to 1985. Price enhancement was achieved by setting floor prices (nonrecourse loan rates), diverting acres from production, and accumulating commodities in storage. Stored commodities only moved to the market if prices were sufficiently above the loan rate to pay for interest and storage or if export subsidies were available to allow the grain to compete on the market at prices below the loan rate.

Since 1985, U.S. policy has changed rather significantly. Perhaps the most important change is that nonrecourse loan rates are no longer fixed but are set below the long run world price by a formula based on the 5-year moving average of market prices. They thus provide a safety net under prices rather than a floor. Because of this change and because deficiency payments are no longer tied to current production at the margin, decisions about production are reflective of market prices. In the absence of export subsidies, U.S. goods would move into the market competitively.

This program change means that the United States is no longer the residual supplier to the international grain market or the seller of last resort. Lower world stock levels will prevail in the future as the United States reduces subsidized storage. The lower level of U.S. stocks may allow greater volatility in world grain supplies and prices in response to yield shocks.

My objective in this chapter is to review specific mechanisms for stabilizing world markets.

Programs considered include: unilateral buffer programs such as the U.S. nonrecourse loan and Farmer Owned Reserve; multilateral grain cartels and commodity agreements; international buffer stock schemes, (including an international organization to hold stocks); a multilateral sharing of responsibility; and country managed yield buffers.

Forces Producing Supply and Price Change

Forces producing supply and price variability include economic forces, seasonal variability, and weather.

Economic Forces

Important sources of change for both supply and demand in agricultural markets include, among other things, macroeconomic conditions or the general economic health of the economy, country monetary and fiscal policies, international flows of capital, trade in nonagricultural markets, changes in exchange rates, and policy actions that restrain or enhance commodity flows.

Changes in these factors produce the economic signals that markets should translate into demand and supply changes. These variables produce real shifts in demand for or supply of agricultural products. In open markets, these changes show the need for more or fewer resources in the sector and commodity prices respond to stimulate or slow output. These are the forces that the calculus of the market should capture in market prices.

Seasonal Variability

Agricultural production is seasonal and an imbalance of output and consumption occurs within a production season. Prices are depressed at harvest when the quantity supplied exceeds quantity demanded and are higher in other periods of the year when quantities supplied are short. Grain consumption tends to be evenly distributed across time; thus, commercial entities will attempt to distribute production across a marketing year so that in a normal year the bins are empty just before the new crop.

Commercial firms have little incentive to carry more than pipeline needs from one season to the next, particularly when a higher cost of acquiring grain can be passed on to the final consumer and a higher revenue can be obtained because of the inelastic demand. Total commercial storage with no government stocks will be equal to planned production plus pipeline carryover.

Weather Variability

Aside from policy changes which may produce unexpected conditions in the economic environment, weather is the major source of uncertainty in agricultural output and, therefore, a source of variability in prices. The national or global impact of weather on the quantity of output for any production season is unknown when a crop is planted. Producers must formulate their planting decisions on some expected normal yield, expected price and expected cost. Resources are committed based on these expectations, but yields often vary sharply from the expected level. Wide yield variations result in large aggregate shifts in quantity produced and in price changes that are magnified by the inelastic aggregate demand. Thus, income from production varies from large positive returns to large losses as the result of weather-induced changes in yields.

Because these yield changes are unplanned, they are a noneconomic adjustment to the production system. That is, producers did not invest or withhold resources to achieve the weather-induced yield changes. Commercial producers plan their stock holding on normal weather and yields because these are the most probable outcome. Profits are not earned by investing in the improbable prospects of crop failures. Because they are irregular and unpredictable, crop failures are not commercial propositions until information becomes available that diminishes the likelihood of an average crop. If crop yields are stochastic (allowing for trends and policy), the most profitable expectation is that yields in any future year will be normal.

Commercial firms will meet profit-maximizing objectives by equating the expected marginal cost of storage with the expected marginal revenue from storage. When it becomes known late in a production season that the crop is likely to be less than expected at planting time, commercial firms may find it profitable to carry over grain into the next marketing season from the previous production year on the expectation that prices will rise sufficiently to cover storage cost and return a profit. Commercial firms have little power to hold larger than normal stocks off the market and carry them over into the next or succeeding seasons. Given expectations of normal crops, storage costs become too large for the private sector. Thus, commercial firms have little incentive to meet inter-year supply stability objectives or societal humanitarian objectives.

Government Intervention

Although inter-year storage is not a commercial proposition, there may be valid reasons for governments to subsidize storage of some grain under market oriented conditions. One reason may be the fundamental problem with weather-induced supply variability. That is, prices are induced to change rather dramatically to allocate available output among consumers in response to the yield shock. Because of the inelasticity of demand, consumers find that, in the absénce of a stocks program, they pay sharply higher prices with rather small reductions in yields. And, producers receive sharply lower prices with small increases in yields.

Food security becomes a primary concern as the very poor find they cannot obtain sufficient food to prevent starvation. Political stability of governments becomes a problem as consumers and producers demand action by governments to bring about stability in food prices. Under the conditions that result from weather variability, governments can intervene in the market to correct for market imperfections and improve market efficiency. By inducing inter-year stability in grain supplied, real changes in supply and demand are made more obvious. Price changes more nearly reflect actual market surplus or scarcity. Livestock production stabilizes and resources are used more efficiently. Whether these benefits outweigh the cost of a public grain storage program may not be easily determined, but it seems likely that a well defined storage program that addresses only the noneconomic shock from weather could provide a net benefit to society.

Options for Market Supply Stability

In the past, many options have been considered and used for government stocks and supply stability programs. The following programs are representative of alternatives used or proposed for international supply stability.

Multilateral Grain Cartels

The International Wheat Agreements were multilateral grain cartels designed to stabilize the quantity and price of wheat in international trade. As with most cartels, the lack of mechanisms to enforce purchase and sale agreements and the drive for self-sufficiency and self-determination by individual countries led to their failure as a market regulating tool. Erosion of market shares for major exporters, excess capacity stimulated by supported prices, and growing stockpiles brought about the end of the agreements and the institution of export subsidies to enforce a competitive position in the market.

Such cartels are the antithesis of market orientation and free trade. They depend on fixed prices, import and export quotas, and production controls. Bureaucratic decisionmaking replaces the market and attempts to regulate the exchange of commodities. The inability of bureaucracies to regulate economic systems without the aid of market prices to provide signals is obvious. And, it is clearly demonstrated by the demise of the economic system in the Eastern Bloc countries.

International Buffer Stocks Schemes

Events of the 1970's caused considerable uncertainty in world grain markets. In 1972, the United States devalued the dollar and provided export subsidies, making U.S. grains an unbelievably good buy. Capitalizing on bargain prices, the U.S.S.R. bought out U.S. wheat stocks in 1972 and 1973. Later, large purchases by India, China, and the U.S.S.R. fueled price increases in an already tight market. The specter of rising prices caused the developing and food-deficit nations to call for buffer stock programs that would protect them from a severe drain on their foreign exchange and from political and economic instability. Developing country importers wanted a buffer stock to be held by an international organization. The organization never was developed because exporters never agreed that any international organization could administer such a stockpile. Agreement never materialized on how such a buffer would be established and maintained. A major problem in finalizing an agreement was the donor countries belief that the objectives of the international stock control organization would be at odds with their interests.

Unilateral Management

Without agreement on a multilateral approach, the United States moved forward with two initiatives in the late 1970's to provide buffers against short supplies and rising prices. These were the Food Security Reserve and the Farmer Owned Reserve (FOR). The Food Security Reserve, motivated by humanitarian interest, was to provide food assistance to poor countries that were suffering a food crisis; those who could not enter the market to obtain grain because they were short of foreign exchange or because they did not have the resources to obtain credit.The FOR was to help farmers carry over supplies for up to 3 years so they could benefit from large crops that were carried over to years when crops were short.

The FOR, and the regular loan program, became the U.S. unilateral buffer stock for the world.In concept, the FOR was to establish a price band by carrying stocks over time, but, with the implementation of the soviet grain embargo, the administration began to use it for price enhancement. It failed for several reasons, one being that the market had excess supplies more often than not and stocks accumulated in the reserve. As administered in the late 1970's and early 1980's, the FOR became a price enhancement program.

Options for Buffer Stocks

In the absence of price enhancement as a farm policy goal, a fundamental objective of government stockholding could be to smooth the delivery of a product to the market, thus avoiding shortages and surpluses. Suggestions for ways of accomplishing this have taken several forms. These include target stock levels, proportional quantity buffers, price-triggered acquisition and dispersal rules, and quantity-triggered rules and risk minimization or consumer surplus maximization rules. Any of these could be applied on a unilateral or multilateral basis.

Fixed Stock Levels

Some have recommended the establishment of a fixed level buffer stock that would be large enough to cover 80 or 90 percent of the expected shortfalls in production. However, a target level of stocks is not desirable because it makes the stock level the objective rather than making the stability of the quantity delivered to the market the objective. Experience with target levels shows that they can exacerbate price changes rather than buffer them.

Ever Normal Supply

An early buffer proposal assumed that annual variability in total production should move into and out of storage providing for an ever-normal granary. However, adding to or releasing from stocks in response to changes in the total quantity produced allows increases in output from planned acreage changes and from productivity increases to enter the stockpile. Such stocks cannot be removed from storage by normal production shortfalls. Such a program captures real or economic shifts in supply and demand maintaining prices above market clearing levels, thus distorting the market.

Proportional Buffer

Establishing a target stock as a percentage of total use would seem reasonable, yet such a scheme suffers from the same problem as a fixed stock level, the stock level becomes the objective replacing the stability of market supply objective.

Yield Triggered Buffer

A stocks program managed by a yield rule would stabilize quantity delivered to the market near the expected level and stabilize prices with minimal interference with the allocation function of prices. Prices would be free to respond to real changes in demand and supply. And, the prices generated would not be clouded by the fog created by yield shocks. Because the yield shock from production is not part of expected production, there can be an optimum acquisition and dispersal program that reacts to current year yield shocks, allowing them to enter and to be released from storage. Under a yield rule, timing of a government storage decision is critical.

Given the inability to exactly forecast yield at planting time, one must base storage decisions on yields estimated at the end of the production season for the current crop. The decision on how much of the crop to store or offer to store must be made just prior to harvest when current year yields and acreage to be harvested are easily estimated. A yield-triggered buffer might acquire the positive amount above projected trend yield times the acres harvested at a percentage of a moving average price and offer these stocks for sale, if the price rose by a percentage above a moving average. Producers could choose whether to sell the yield increment to the market or to the government. Only an amount equivalent to the deviation above the national trend times the acres harvested would be permitted to enter government storage from the current year's crop. This would reduce price variation due to yield changes. Figure 7.1 shows U.S. wheat surplus and deficits from expected yield based on a moving trend yield rule identified as the yield shock. The shock is based on actual yields and acres harvested using a 15-year moving trend yield projected 1 year as a basis for establishing expectations. Actual stocks of wheat are shown as the line identified as stocks on the chart. The potential stocks line assumes that all positive deviation above the expected were purchased and all negative shocks were dispersed when stocks were available. The data are illustrative of the possible quantities of wheat purchases and sales under a yield buffer arrangement. Assumed purchases varied considerably but stayed below 5 million metric tons per year. Total potential stocks rose to about 25 million metric tons and declined to zero.

Figure 7.2 shows wheat stock data for the world. World stocks exceed yield shock adjustments by a considerable margin which suggest that food security may play a substantial role in the decision of countries like India, Pakistan, and China to hold stocks. To a large extent, the build-up of stocks from 1981 to 1986 was the result of U.S. farm policy.

Figure 7.3 shows data for U.S. corn stocks, potential stocks and production deviations based on yield shocks. The effects of U.S. price support policies on stocks can be seen rather clearly. Stocks grew rapidly under the 1981 Agriculture Act. Weather and the Payment-in-kind (PIK) program caused stocks

112

Figure 7.1 U.S. wheat stocks and potential stocks from yield deviations
Million metric tons

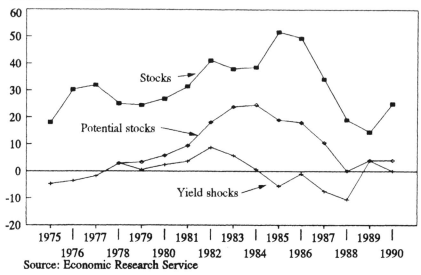

Source: Economic Research Service

Figure 7.2 World wheat stocks and potential stocks from yield shocks
Metric Tons

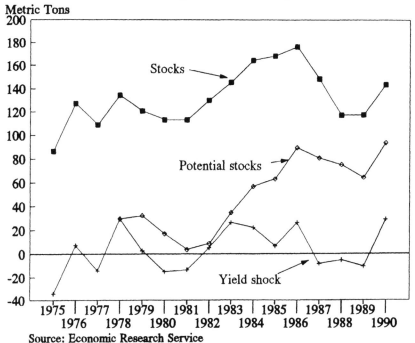

Source: Economic Research Service

Figure 7.3 U.S. corn stocks, potential stocks and deviations from yield shocks
Million metric tons

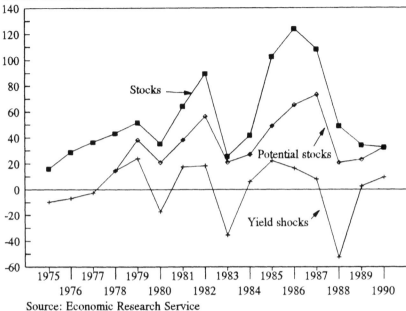

Source: Economic Research Service

Figure 7.4 World corn stocks, potential stocks an yield shocks
Million metric tons

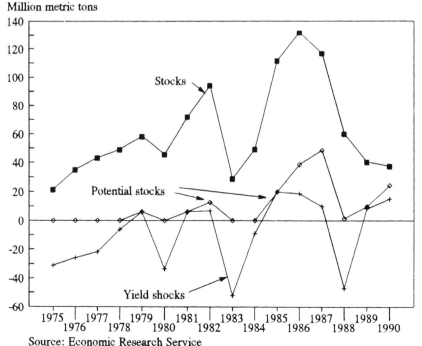

Source: Economic Research Service

to decline, but they again grew rapidly until the implementation of the 1985 act. Since 1986, U.S. policy has caused stocks to decline sharply.

Figure 7.4 shows corn stock data for the world, excluding China. The data for China were excluded, because they do not seem to produce results that are compatible with the estimation procedure used. Stocks for China increased continuously and then declined without demonstrating stochastic properties. The data in figure 7.4 suggest that, in the absence of government price enhancement storage programs, yield variability would have emptied the storage bins twice during the 1980's. However, there is considerable interaction between corn and other feed grains and a fairly elastic demand for corn relative to wheat. Thus, the price effects may not have been as severe as they would be with wheat. Storing more than the positive deviations from trend would mean storing more than had been planned for by producers or expected by consumers. Storing less than the positive deviations would mean that the probability of incurring a shortfall in stocks increased. If long-term supply and demand were in balance, the smoothing effect of the yield storage rule on quantity would result in a stability of domestic consumption and a stable supply for export.

Summary

The objective of an independent stocks program would be to smooth the flow of unanticipated product to the market in response to unanticipated short supplies. If resources are committed with the expectation of normal yields and prices and the output results in a significantly better or poorer crop, prices and incomes can be dramatically altered, even though the producers planned appropriately given their limited information. Neither the Government nor the farmer can correctly anticipate or forecast the outcome of a specific crop at planting time except by chance. Stocks programs, therefore, should react to crop output rather than anticipate crop output. Protecting farmers and consumers against random shocks to the system need not distort long-term market signals if the shocks are due to weather.

References

Farnsworth, Helen C., "International Wheat Agreements and Problems," 1949-56, *Quarterly Journal of Economics*, pp. 217-248.Foreign Agricultural Service, *International Grains Arrangement 1967*, FAS-M-195, U.S. Department of Agriculture, Washington, DC, Nov. 1967.

Hillman, Jimmye, D., D.Gale Johnson and Roger Gray, *Food Reserve Policies for World Food Security*, Rome, U.N. Food and Agriculture Organization, 1975.

Johnson, D. Gale, and Daniel Sumner, An Optimization Approach to Grain Reserves for Developing Countries, *Analysis of Grain Reserves, A Proceedings* ERS-634, Economic Research Service, U.S. Department of Agriculture, Aug. 1976. pp. 56-75.

United States Senate, *International Commodity Agreements: A Report of the U.S. International Trade Commission*, Washington, DC: Committee Print of the Committee on Finance, 1975.